THE HISTORY
OF FINNISH LITERARY CRITICISM
1828 – 1918

The History of Learning and Science in Finland
1828 – 1918

15 a

Editorial Note

The history of learning in Finland could be said to begin with the foundation in 1640 of the Academia Aboensis at Turku (Åbo). Finland was at that time a part of the kingdom of Sweden. As a consequence of the Napoleonic wars the country became an autonomous Grand Duchy under Russia. In 1828, after Turku had been devastated by fire, the university was moved to the country's new capital, Helsinki (Helsingfors), and its name was changed to the Imperial Alexander University of Finland. It became Helsinki University in 1918, when Finland emerged as an independent republic from the turmoil of World War I.

The history of learning in Finland in the period 1640–1828 has been recorded in the ten volumes of the series Åbo Universitets Lärdomshistoria (The History of Learning in the University of Turku). The series was published round the turn of the century by the Swedish Society of Literature in Finland. It remains the standard work on the early history of scholarship and science in this country.

The idea of treating the history of learning in Finland during the time of the Imperial Alexander University in a series of monographs was first suggested by the late Chancellor of Helsinki University, Edwin Linkomies. The idea was followed up by the Finnish Society of Sciences, which announced its realization in connection with the Society's 125 th anniversary in 1963.

During the whole of its existence, the Imperial Alexander University was the only university in Finland. Practically all research and scholarship during the period 1828–1918 was therefore intimately connected with this one institution of learning and higher education. This, in combination with the facts of political history, gives to the material presented in the projected volumes a historical unity which has been thought to justify their planning and publication as a series.

According to the editors' plans, the series will comprise 20 volumes, some of which will be divided into two or more self-contained parts. The authors are distinguished representatives of the various fields of scholarship or science, with the history of which their conmtributions deal. In practically all cases the ground covered has not been treated in a monograph before. The works are thus the fruit of original research on the part of their authors in the history of ideas and learning.

THE HISTORY OF FINNISH LITERARY CRITICISM 1828 – 1918

by
Yrjö Varpio
Professor of Finnish Literature
University of Tampere

Tampere 1990

Translated from the Finnish manuscript by
Douglas Robinson

Photos reproduced with the kind permission of
the National Board of Antiquities (M)
the Finnish Literature Society (SKS)
the Swedish Literature Society in Finland (SLF)
the Helsinki University Library (HYK)
the Helsinki City Museum (HKM)
the Åbo Academy Library (ÅAB)

ISBN 951-653-212-8

Printed in Finland by Hermes
Tampere 1990

Contents

Preface

The history of Finnish literary criticism between 1828 and 1918 is largely the history of scholarship at the University of Helsinki. After the fire in Turku, the university was moved in 1828 to Helsinki, and it was not until the end of the period under study that two more universities were founded, both in Turku. In my account of this history I have, however, looked at writers from outside the university as well, since the most significant literary historians, critics, and essayists without academic appointments have played an important role in debates over the goals of literary criticism, and thus have offered an important challenge to academic criticism. My primary emphasis remains, of course, on the academic world.

I began writing this book in 1986, when the Finnish Society of Sciences project for the writing of the history of various branches of Finnish scholarship was already well advanced, and several volumes in the series had already appeared. I had just completed an overview of Finnish literary criticism in Finnish that extended into the present, and that book served as my groundwork for the present study. It was also helpful that the University of Tampere granted me funding to compile the first bibliography of nineteenth-century Finnish literary criticism. The relative youth of the University of Tampere Library caused some difficulties, since most of my primary sources had to be obtained on interlibrary loan from Helsinki and Turku; the librarians at the University of Tampere have been most helpful. I also received invaluable assistance from the staff at the University of Helsinki library and archives.

My research was facilitated by many conversations with Päivi Huuhtanen on the gray area between literary criticism and aesthetics. Ilpo Tiitinen, an excellent scholar of the history of learning materials in Finnish literary criticism, read the manuscript and offered many helpful suggestions. Per-Erik Cederholm helped me interpret old Swedish terms. The editor-in-chief of the series, Professor Georg Henrik von Wright, has been patient and understanding with my slow progress toward completion. Both he and Kauko Pirinen read the manuscript and made many invaluable suggestions. The book's current form is in large part the work of Professor Douglas Robinson, who translated it into English. Thanks to all who helped make the book what it is.

Tampere, May 18, 1990
Yrjö Varpio

Translator's Note

Translators rarely get the chance to work in as many of their fields of expertise at once as I have here: as a literary critic, historian, and theorist, and as a translator and translation theorist I felt more qualified to undertake the translation of Professor Varpio's book on the history of Finnish literary criticism than I have in my fifteen years of translating.

But I found, once I started on the translation, that the work was no easier for all my expertise and experience. The study of literature, like all subjects and methods of human inquiry, has developed differently in different countries. As an American who has studied, taught, and written about literature for fourteen years in Finland, but whose professional center of gravity has nevertheless always been the United States, these differences have been part of my academic life for a long time, whenever a conversation stumbled over words like *kirjallisuustiede*, which we Americans would rather not render the "science of literature," or *literary criticism*, which Finns always understand to be the journalistic practice of reviewing books. But in this translation such stumblings became a way of life.

I have rendered *kirjallisuustiede* (literary science) and *kirjallisuudentutkimus* (literary research) variously, depending on my sense of the context, as literary criticism, literary scholarship, literary studies, or literary science (or the science of literature). Ever since Matthew Arnold ennobled literary criticism in his influential 1864 essay on "The Function of Criticism at the Present Time," that has been the central term in the Anglo-American academic world for what academic professionals do with literary texts. It does not imply reviewing; it is the interpretation of literary texts, usually but not exclusively literary classics, by academically trained university teachers.

In comparison with literary scholarship, to be sure, criticism does have a slightly amateurish ring – until fairly recently, not all critics had Ph.D.'s, and criticism was understood to be something any highly intelligent and fairly well educated person could do – but in the Anglo-American world, as in fact in Finland as well, that amateurish taint was never considered particularly damning. Even today, when criticism is highly professional, the critic is conceived as someone whose concern is less with abstruse *truth* than with cultural coherence and tradition – with legislating for cultural knowledge, as Arnold would say.

Literary scholarship is sometimes taken to cover the entire field of what is more commonly called literary studies – all approaches to literary texts – but in its technical sense it is specifically a rigorous philological discipline concerned with the development at the microlevel of a given word, motif, or myth across an entire period, or with the textual variants in a given author's editions, or the like. Scholarship in the Anglo-American world is associated with the Germanic tradition, especially people like E. R. Curtius and Erich Auerbach, and it carries a good deal of prestige; but it is not commonly practiced.

Kirjallisuustiede and *kirjallisuudentutkimus* are used almost synonymously in Finland. But the Germanic term literary science, or the science of literature (Literaturwissenschaft), has a phony feel to it in English. It sounds like humanists pretending to the methodological rigor of the physical sciences, in an insecure longing for the social status of the Scientist. When it is used in Anglo-American academic discourse at all, it is associated less with nineteenth-century positivism, where its ideological roots are, than with French structuralism of the 1960's, where it formed part of the project of the so-called "human sciences." In the translation I use the phrase as infrequently as possible – and then only when the critics in question (particularly toward the end of the nineteenth century) patently see themselves as striving for "hard" scientific rigor in the analysis of literary texts.

Another problem lies in the translation of various Finnish institutional practices. Such Finnish terms as *virantäyttö* (appointment), *pätevöitymisaika* (extension on an application deadline to enable a candidate to finish writing a book or article that will weigh in the appointment process), *asiantuntijalausunto* (peer review statement), *vastaväittäjä* (official faculty opponent), and *äänestyslausunto* (voting statement) all ground my translation in a nebulous middle region halfway between Finnish and American academia. The rigidity of the Finnish appointment process, for example, generates words for legislated flexibility like *pätevöitymisaika*, which is built into the American appointment process, where the tenure track gives appointees what amounts to a five-year "extension" on the application deadline. In Finland peer review is undergone before the appointment; in America only after the appointment, for tenure and promotion. In America doctoral dissertations are evaluated by a reading committee, in a closed seminar situation; in Finland dissertation defenses are major public events presided over by a *kustos* (which I have rendered "chair") and the *vastaväittäjä* (whom I have rendered "official faculty opponent"). These are the sort of *realia* that make most translation difficult; I hope these prefatory remarks will help the reader to navigate through the intermediary waters of my renderings.

I have in fact used American English throughout the translation, and have, as the above examples make clear, situated Finnish institutional practices in an explicitly American translational context – simply because I am an American and am most familiar with American academia. For any inconvenience this may cause British or other English-speaking readers, I apologize. I believe it is preferable to translate into a living linguistic tradition, one the translator knows and feels and cares about deeply, than to attempt to construct a carefully denatured laboratory language, a kind of translational Esperanto, for fear of using words or phrases that feel too regional or parochial to some readers. If this translation works, it will capture my specifically *American* sense of what it feels like to teach and do scholarship and negotiate the institutional ins and outs of a Finnish university.

A note on documentation. I have used the Modern Language Association style throughout, with parenthetical references in the text to items listed in the bibliography. In each case the minimal information required to help the reader locate the item in the bibliography and the pages therein is provided within parentheses. If the source is clear without parenthetical reference, none is given. Notes are reserved for lengthier explanations.

INTRODUCTION
Review of Secondary Sources

The historical study of Finnish literary criticism is a fairly recent phenomenon. It did not get underway until the 1970s, and then largely as part of an international trend: around that time Swedish, German, and Anglo-American scholars became increasingly interested in the history of criticism, and the Finns followed suit.

The most comprehensive study of the subject is my own 1986 academic textbook, *Suomalaisen kirjallisuudentutkimuksen historia* ("The History of Finnish Literary Criticism"), on which this book is based; there I traced the development of scholarly approaches and the critics who propounded them from the early nineteenth century to the present. That book superseded Rafael Koskimies's brief overview of the subject, "Kirjallisuuden tutkimus" ("Literary Criticism"), in the seventh volume of the 1968 work *Suomen kirjallisuus* ("Finnish Literature"). In his *Maailmankirjallisuuden yleisesitykset Suomessa* ("Finnish Surveys of World Literature," 1989) Hannu Riikonen studies Finnish histories of world literature specifically in terms of the literary-historical conceptions on which they are based and of the position assigned to various national literatures and authors at various times. Narrower studies would include E. N. Tigerstedt's article "Finländsk litteraturvetenskap 1914-1938" ("Finnish Literary Science," from 1939); Olof Mustelin's article "Från lärdomshistoria till litteraturhistoria" ("From the History of Learning to the History of Literature," from 1971), which focuses on the early nineteenth century; Kai Laitinen's article "Kirjallisuushistoria: linjanvetoa ja ongelmia" ("Literary Histories: Demarcations and Problems," from 1980); and my own article "Die finnische Literaturforschung in den letzten zehn Jahren" ("Finnish Literary Criticism over the Past Ten Years," from 1980). Arvid Hultin's excellent series of studies done between 1914 and 1918 on the early years of the University of Finland (1640-1828) should also be mentioned, as should I. A. Heikel's volume *Filologin* ("Philology," 1894) in his study of the history of learning at the University of Turku.

A number of works written in the 1970s on other subjects also dealt in passing with the history of Finnish literary criticism. Pertti Karkama's book *Metodi ja maailmankatsomus* ("Method and World View," 1974) was an attempt to demonstrate how the choice of an analytical method is always an ideological choice, and in the course of his argument he made important points on the positivistic science of literature in Finland in the early twentieth century. Päivi Huuhtanen's book *Tunteesta henkeen* ("From Feeling to Spirit," 1979) was centrally concerned with the development of Finnish aesthetic theory during the first third of this century, but in demarcating aesthetics she also traced the theoretical backgrounds of the period's literary criticism. In the Finnish Society of Sciences series on "The History of Learning and Science in Finland 1828-1918," Sixten Ringbom's volume on *Art History Before 1920* (1986) and Pentti Aalto's volume *Modern Language Studies in Finland 1828-1918* (1987) both deal in passing with the work of literary critics, because until the end of the nineteenth century both of these fields (art history even longer) formed a part of the broad discipline entrusted to the Professor of Aesthetics and Literature at the University of Helsinki. Other volumes in the same series, notably Jouko Hautala's *Finnish Folklore Research 1828-1918* (1968) and Mikko Korhonen's *Finno-Ugrian Language Studies in Finland 1828-1918* (1986), also offer interesting insights on the subject, since folklore studies during that period were carving out a niche of their own in the gray area between the chairs of Finnish language and literature, and since the study of Finno-Ugric languages included teaching in Finnish literary history as well.

Biographies of several literary critics during the autonomous period have been written: Fredrik Cygnaeus, August Ahlqvist, C. G. Estlander, Julius Krohn, Werner Söderhjelm, Yrjö Hirn, and Viljo Tarkiainen.[1] Other central biographies, however, remain unwritten: Valfrid Vasenius, Eliel Aspelin-Haapkylä, and Gunnar Castrén. While biographies are by definition more concerned with individuals than with large-scale cultural developments, they are highly useful sources of information about the cultural activities in which their subjects were engaged, and I have made extensive use of the biographies that are available.[2]

Finnish methodological debates are often carried on and recorded in published inaugural addresses by full professors of literature and papers at scholarly conferences. Additional information on specific figures can be obtained in orations and obituaries delivered and written about them after their death. In this latter case the uncritically laudatory nature of the remarks forces the historian to be particularly cautious. Another and more reliable source is the external peer review written during an appointment process; in the Nordic countries these are typically extensive and detailed evaluations of the candidates' scholarly merit. If we read these peer review statements as historical responses to the scholars in question, we can determine not only what the scholars were doing but how their activities were perceived in their time: what was considered old-fashioned or innovative, what desirable or undesirable.

Sulo Haltsonen's *Luettelo suomalaisista kirjallisuudentutkimuksista 1901-1925* ("Index of Finnish Literary Research," 1954) and Marja Lehtinen's *Luettelo suomalaisista kirjallisuudentutkimuksista 1801-1900* ("Index of Finnish Literary Research," 1989) contain useful information about literary-critical books, articles, and (selectively) also newspaper articles published in Finland. Also included in these indexes are Finnish studies published outside Finland and foreign studies of Finnish literature. The indexes make clear that while the major language of Finnish literary criticism until the end of the nineteenth century was Swedish, Finnish-language criticism was also beginning to be published in increasing quantities at the beginning of the twentieth century, mainly in article form. Doctoral dissertations were published in Latin until the mid-nineteenth century; in 1852 Swedish and in 1856 Finnish were approved as dissertation languages. If however one checks the original languages of the literary-critical books published between 1853 and 1918 by critics with Ph.D.s, Swedish is far and away the dominant language.[3]

	1853–1900	1901–1918
Swedish	75	40
Finnish[4]	7	20
French	4	5
German	6	1
English	1	–
Italian	1	1

PART ONE
Background

Early Years at the University of Finland, 1640-1828

The roots of Finnish literary criticism extend back to the founding of the Academia Aboensis in 1640.[5] Even in the seventeenth century certain professorships in the humanities covered pedagogical and scholarly ground that might loosely be described as literary-critical, and as the history of learning took hold at the Academy in the eighteenth century, teaching and research in literary history, poetics, and aesthetics were increasingly actively pursued. The analysis of fictional texts served quite a different function, however, than it would in later centuries. To be sure, the professor of Greek and Oriental languages dealt with worldly writers as well as the Bible, but the main emphasis was on language teaching. The central duties of the professor of eloquence (Latin) were to teach the language and to do rhetorical analyses of famous speeches. In fact the first editions of ancient Greek and Roman classics published in Finland were either collections of orations or historiography. Bishop Johannes Gezelius's press in Turku published a selection of Cicero's speeches and letters (1669), Cornelius Nepos's *De excellentibus ducibus exterarum gentium* (1674), Florus's history of Rome (1675), and Curtius Rufus's *Historiae Alexandri Magni Macedonis* (1675). Emphasis was placed on nonfictional works precisely because the objective of instruction was a command of Latin, both in writing (reading and producing) and in speaking (oratory). Literary texts were perceived as of secondary importance.

Rhetoric was the central theoretical turf of Latinists at the University of Turku, and most professors of eloquence published studies of it. One of these was the most noted Finnish humanist of the seventeenth century, Enevaldus Svenonius (1617-1688), who wrote the first (if brief) Finnish history of Roman literature, indeed the first Finnish literary-historical book, *De lingvae latinae puritate.*[6] Svenonius covers the twenty most significant Roman writers (including, to be sure, more historians and orators than poets) with short biographies, bibliographies, and summaries of the best-known commentators' appraisals.

Since the academic practice of the day required students to be able to imitate classical models in the writing of laudatory and elegiac poems in Latin or Greek, in 1655 the first professorship of poetry was instituted at the

Academy. The professor's duties included teaching students versification according to the poetic principles taught by Aristotle and Horace and exemplified by Horace, Virgil, and Ovid, and checking the correctness of poems for publication. That the emphasis in all this was on formal linguistic and poetic correctness is evidenced by frequent complaints in Academy documents that a student had been allowed to publish a formally inadequate poem without advance screening.

The most interesting documents of the late seventeenth century from a literary-theoretical point of view are the statements in which professors of poetry defended their subject in the generally antipoetic climate of the Academy. David Lund, for example, who served as professor of poetry in the 1690s, defended poetry on the grounds that it taught not only morality, which was generally accepted as its task, but also the universally useful skills of articulation and style.[7] Other apologias were offered as well; in 1704 the professorship of poetry was considered essential "because the Finnish nation was fond of poetry and it was therefore important that they have someone to guide them in it" (Heikel 108).

It goes without saying that the classic poetics of Aristotle and Horace and others were drummed into students of philology in the seventeenth and eighteenth centuries. In the late seventeenth century a new approach to these and other ancient texts began to take hold: such texts were not merely of instrumental value, in the transmission of sacred texts and scholarship, but of considerable scholarly interest in their own right. Henrik Hassel (1700-1776) and his successor as professor of eloquence Henrik Gabriel Porthan (1739-1804) developed this enlightened notion still further by linking the cultural heritage of Greece and Rome with contemporary needs, particularly the development of culture in the native tongue. Porthan's immense significance in the early history of Finnish scholarship arises out of precisely this watershed or transitional work: he linked an interest in classical poetry with an interest in Finnish folklore, for example, and advised his doctoral students to apply classical rhetoric to contemporary analytical needs, such as the study of sermons.[8]

While most doctoral dissertations of the day demonstrate far more knowledge of sources, respect for authorities, command of ancient languages, and industry than logical inference or originality, some dissertations from the prehistory of Finnish literary criticism are worth mentioning. These would include the Swedish-born professor of poetry Petrus Laurbecchius's (1628-1705) dissertations on poetics *In Aristotelis Poeticen commentarius* (1674, the first Finnish study of poetics, which apart from a few pages was destroyed by fire later in the seventeenth century) and *De artis poeticae existentia, nomenclatura, genere et materia* (1680); Petrus Malmström's *Specimen poeseos Pindaricae I* (1789), the first professorial dissertation to deal with a secular author; and Per Johan Alopaeus's (1773-1814) *Specimen litterariae Fennicae* (1793-1795), a four-volume work that compiled bibliographical information about the writings of primarily Academy teachers. In addition numerous dissertations written under the direction of Torsten Rudeen (1661-1729, professor of poetry), Hassel, and Porthan dealt with issues that can be considered literary-theoretical.[9] In the area of aesthetics one popular foreign authority in the late eighteenth century was the German Johann Georg Sulzer, whose comprehensive index of aesthetics offered a highly accessible source for early Finnish students of the subject.[10]

Characteristic of literary studies in the University of Turku era was above all their close linkage with other disciplines and their instrumentalization as a means to students' general cultural competence. Porthan's inaugural address upon his appointment to the professorship of eloquence in 1777 is in this sense typical: he ponders the utility of poetry for the proper development of other disciplines (see Heikel, 223). The generality of literary studies in the period is also evident in the disciplinary mobility of scholars: professors did not specialize in a restricted field, but moved freely from chair to chair in the general direction of the highly prestigious professorships of theology. A representative case is Laurbecchius, who was appointed professor of poetry after he had served several years as professor of mathematics, and as professor of poetry lectured not only on poetics but on jurisprudence and theology as well, and ultimately was appointed professor of theology.

At the University of Turku teachers and students came in contact with roughly two kinds of literature: the "old" texts of the Bible and classical antiquity, and the "new" texts of contemporary occasional poetry. The contemporary national literature that from a later point of view was the most significant of the era, oral folk poetry, was almost entirely unknown.

In her dissertation on awareness of Finnish folk poetry in the literature of the sixteenth through eighteenth centuries, Annamari Sarajas explores the main events in the development of an academic cognizance of native folklore.[11] According to Sarajas folk poetry was not studied academically until Porthan's time in the late eighteenth century. Prior to that era folk poetry was written down more or less haphazardly, and the results were used primarily for other purposes than the study of poetry per se. The scanty analytical data available at the time on Finnish folk poetry were summed up by the Swedish scholar Anders Bure in the explanations to his map *Orbis Arctoi Nova Et Accurata Delineatio Auctore Andrea Bureo Sueco* (1626): "In addition the Finns do not use rhyme, as do other nations, but rather alliteration extending over two or three words; and they use an octosyllabic line" (62).

A few seventeeth-century dictionaries and grammars printed samples of Finnish folk poetry in the *Kalevala* verse form (trochaic tetrameter), and in the centuries of Swedish hegemony folk poetry was occasionally assigned the lofty task of projecting a glorious past for the nation. The Church took a dim view of folk poetry; indeed a large portion of the recorded folk poetry in the early centuries of the modern period appears in the court records of witch trials.

Beginning in the 1760s in Turku two separate methodological approaches to the old folk poems took hold. One remained practical, and specifically historical: by studying Finnish language and folk poetry scholars attempted to determine the origins and early lifestyle of the Finnish people. In addition, however, the first steps were taken toward a literary-critical approach to these texts. Significantly, it is in this period that in Finland – as in Sweden and Germany – it became acceptable to study the literature of one's own country. Porthan initiated the study of old Finnish folk poems in his classic work *De Poësi Fennica* (1766-1778) just two years after the Swedish scholar J. H. Lidén had published his *Historiola Litteraria Poetarum Svecanorum* (1764), the

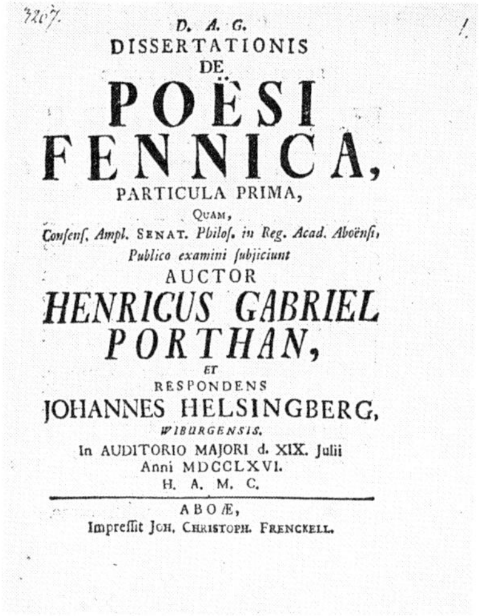

D. A. G.
DISSERTATIONIS
DE
POËSI
FENNICA,
PARTICULA PRIMA,
QUAM,
Confenf. Ampl. SENAT. Philof. in Reg. Acad. Aboënfi,
Publico examini fubjiciunt
AUCTOR
HENRICUS GABRIEL
PORTHAN,
ET
RESPONDENS
JOHANNES HELSINGBERG,
WIBURGENSIS,
In AUDITORIO MAJORI d. XIX. Julii
Anni MDCCLXVI.
H. A. M. C.

ABOÆ,
Impreffit JOH. CHRISTOPH. FRENCKELL.

Henrik Gabriel Porthan and the first volume of his dissertation *De Poësi Fennica* (1766-1778). "I consider it a scandal not only that a native-born Finn does not know our poetry, but that he does not admire it." (M)

national literary history of Sweden. But in fact Porthan's many-volume dissertation differed structurally from Lidén's work, and it is impossible to speak of direct imitation. Whereas Lidén studied his poets biographically, Porthan attempted to analyze actual poems. He included several "high-cultural" or "art" poets, but all of the poems he discussed shared what he dubbed and analyzed as the "Finnish verse form" – which was in fact, in the first volume, the focus of his study. In later volumes of the dissertation, appearing in the 1770s, Porthan specifically addresses the question of folk poetry. This is, of course, part of a general European trend, the scholarly "discovery" of folk poetry. The paucity of recorded oral folk poems complicated Porthan's work, but despite his difficulties he did succeed in offering numerous specific definitions of syllabic length, stress, alliteration, and repetition, all prosodic questions that were to become central to Finnish folkloristics in the next century.

Porthan's work is not yet characterized by that romantic enthusiasm and admiration for the national spirit that was to motivate Finnish students of folk poetry after Finland was split off from Sweden and became an autonomous grand duchy of Russia in 1809. Even before the work of Elias Lönnrot in the compilation of the *Kalevala* in the 1830s and 40s, romantically minded scholars had proclaimed that folk poetry is in fact the backbone of all national literature (A. I. Arwidsson), and that a work could be compiled out of the old folk poems through which "Finnish nationality would awaken by the glory and honor of its own unique nature" the admiration of both contemporaries and posterity (K. A. Gottlund).[12]

In the mid-eighteenth century the professorship in poetry at the University of Turku was changed to a professorship in economics; this was considered better suited to the needs of the time.[13] Around the same time, however, a new discipline was introduced to the Academy, the history of learning, historia litteraria, out of which developed, by the mid-nineteenth century, the first literary-critical discipline in Finland: aesthetics and modern literature.[14]

The History of Learning as the Precursor of Literary Criticism

Beginning in the mid-eighteenth century the history of learning was a popular, even fashionable discipline, and Finnish academics followed its international developments closely. There was some disagreement at the University of Turku as to the degree to which the study of literature should be incorporated into the history of learning; Porthan was one of those who believed it formed an integral part of the discipline.

In the last few decades of the eighteenth century that branch of the history of learning that dealt with literature underwent significant changes. The earlier antiquarian approach to literary texts, with its emphasis on collecting, listing, and grouping, was increasingly displaced by a new attention to the larger cultural-historical background of literary developments and a new tendency to conceive poetry and generally literature as a field in its own right.

This new approach was powerfully reinforced and consolidated by Porthan's multivolume dissertation, and it was apparently due to the interest aroused by Porthan's massive work that the first docent in Finnish literature was appointed at the Academy in 1768. The appointee was Carl Gustaf Weman (1740-1803), who had written for the docentship a modest dissertation whose premises were later discredited, showing the philological congruences between Finnish and Hebrew. Weman only taught at the Academy for a few years; in the fall of 1768 we know that he lectured on the nature of the Finnish language in prose and verse, and in the 1771-1772 academic year on the principles of poetry, especially Finnish poetry. A few years later Weman left the Academy to become a Lutheran minister.[15]

Two other docents of the era are also worth mentioning. In the 1790s, docent in learning history Per Johan Alopaeus gave one lecture course on the literary history of classical antiquity and another on the literary history of the medieval and modern periods.[16] The poet Frans Mikael Franzén (1772-1847), who was docent of eloquence for a few years in the 1790s before becoming professor of the history of learning, lectured on the theory and history of literature while a docent and dedicated himself entirely to literary and aesthetic matters in the professorship. Franzén was the greatest Finnish poet around the turn of the century; in 1811 he moved to Sweden, where he served first as a vicar, then as bishop of Härnösand.

The first professorship in the history of learning was not established until Russian rule: in the 1811 and again in the 1828 statutes the university librarian's post was given an expanded definition so as to include a professorship in the history of learning. First in Turku, then, in 1828, when the University became the Imperial Alexander University of Helsinki, this post was held by Fredrik Wilhelm Pipping (1783-1868, from 1814 to 1838), Alexander Blomqvist (1796-1848, from 1838 to 1848), and Carl Wilhelm Törnegren (1817-1860, from 1851 to 1853).[17] All three also served as university librarian, though in rather complex permutations: Pipping remained in the librarianship until 1845, and Blomqvist only served in the combined post for three years. When the post was split in 1852, its last incumbent, Törnegren, continued on as university librarian.

The general trend in this period was for the role of literary history in the history of learning to become more significant as 1852 approached. Of the three professors of the history of learning, Pipping was the least interested in the history of literature. He had in fact written his Master's thesis on the *Iliad* (1805) and his dissertation for a docentship in Greek literature on Homer's use of epithets for Apollo (*Epitheta Apollinis Homerica brevi explicatione dilucidata*, 1806). There was a powerful reawakening of interest in Homer at the time in German universities, and Pipping's work in the field must be tied to that trend. He was primarily a librarian, however, and in particular a pioneering bibliographer; as professor of the history of learning Pipping placed particular emphasis on bibliography and the systematization of learning.[18] He also stressed the importance of historical thought, however: we know that in his early lectures he insisted that literary culture be examined against the backdrop of the external circumstances of the period, the "political upheavals, the religious changes, and the governing principles" of the day.[19]

Pipping's successor, Alexander Blomqvist, was a cut-and-dried literary historian.[20] Upon publication of his 27-page dissertation on Greek tragedy, *Quae ab Aeschylo accesserunt momenta tragoediae Graecae, secundum Aristotelem in libris poëtices c. IV* (1823), he was appointed docent of Greek literature. He then set about qualifying himself for the chair in the history of learning, which he understood as fundamentally the history of literature. This is clear in the letter he wrote to his friend A. J. Sjögren in the fall of 1823: "My research seems . . . to be turning more and more centrally to drama, which branch of poetry is undoubtedly the greatest, as it is ever the best and surest index of a nation's highest development, in both the moral and the aesthetic spheres" (September 27, 1823). Or again, a few months later:

> The history of literature in its profoundest sense, which is to say its world-historical rather than bibliographical aspect, or what I would want to call in the broadest sense aesthetics, has always interested me and continues to do so. . . . I would say that art in its totality, but above all poetry, along with theology and philosophy, are the channels through which the human spirit has been able to demonstrate its power to act without other leaders than itself and its own good genius. Only thus do I believe it possible for us to attain clarity regarding our own being. (Letter to Sjögren, January 21, 1824)

In this second letter Blomqvist mused that, unless university studies were enlivened by a "philosophical, theological, or aesthetic" principle, they would remain superficial and would have nothing of the universal about them.

In the 1820s Blomqvist left the university for a while. In 1825 he became

---

The Senate Square in Helsinki and the downtown area designed by C. L. Engel after F. Tengström's lithograph in 1838. On the left the Imperial Alexander University, which was completed in 1832. In the middle background the university library, construction on which was begun in 1836 and completed in 1845. (M)

editor of the official government organ, *Finlands Allmänna Tidning*, and in fact served in this capacity alongside his university career until his death. When a public library was founded in Helsinki in 1826 to serve the needs of the bureaucracy, Blomqvist was appointed librarian. Two years later, when the university was moved from Turku to Helsinki, Blomqvist's library was merged with the university library and Blomqvist became an associate professor in the history of learning and assistant head librarian.

In the 78-page dissertation he published for the professorship in the history of learning, *De principio methodi ethnographicae in historia litteraria universali adhibendae schediasma* (1838), Blomqvist traces the development of scholarship in the history of learning and explores the nature of the discipline. *Historia litteraria* is, he claims, "the study of the history of intellectual life as it has manifested itself in science and the verbal arts." Among his numerous sources, Hegel's lectures on the philosophy of history and the history of philosophy and particularly the work of Wachler are crucial. He agrees with Wachler on the application of ethnographic or technological methods, conceived synthetically or chronologically, to the history of learning. He personally favors the use of the ethnographic method. Nations, he says, like their languages and literatures, develop in periods. A poetic or religious period is followed by a prosaic or philosophical period, during which philosophy as the medium of national consciousness assumes the function earlier performed by a mythically conceived religion-poetry axis. In this phase poetry becomes subjective, shifting from epic to lyric and culminating at its apex in the drama. Following Wachler, Blomqvist argues that the ethnographic method is best suited to the study of the history of learning because it can shed the greatest light on the most fascinating focus of historical inquiry, "folk nature itself, the people's intellectual life."

During Blomqvist's tenure the actual content of the history of learning changed dramatically. In the spring of 1840 Blomqvist wrote his friend Sjögren of his "great plan" to shift the traditional history-of-learning topics to his lectures on aesthetics (May 24, 1840). In the fall of the same year he began his lecture course on "literary aesthetics." The written text of his lectures has survived, and they show clearly how the Finnish history of learning was already moving in the direction of literary criticism and aesthetics. Between 1834 and 1848 his lecture courses included an introduction to literary history (i.e., the history of learning) and bibliography (1834), the history of literature (1834, 1835), the history of Greek literature (1835), bibliography (1836, 1837, 1838), general literary history (1837), the history of medieval literature (1838), general literary history (1838, 1839, 1840), literary aesthetics (1840, 1841, 1842), the history of modern literature (1840), the history of dramatic literature (1843), the history of modern poetry (1843), the history of poetry (1844), aesthetics (1844, 1845, 1846), an introduction to literary history (1846), and general literary history (1848). Blomqvist's lectures were significant not only in his choice of topics, but in his application of the philosophy of Hegel. Hegelianism was influential in other sectors of Finnish thought of the day; Blomqvist was the first Finnish scholar to apply it to the study of literature and aesthetics.[21]

In the final phase of the history of learning many young scholars doing their dissertations in the field turned toward the history of literature and aesthetics, among them Carl Wilhelm Törnegren, Fredrik Berndtson, Gustaf Edvard Ingelius, Johan Oskar Immanuel Rancken, and Theodor Clasen. Of these Berndtson and Clasen were appointed docents in the new field of literary criticism, Törnegren and Ingelius performed their life's work in the university library, and Rancken taught high school. Since Törnegren's scholarly work ended while he was still professor of the history of learning, this may be an appropriate place to examine it.

Törnegren had taken his Master's degree at the University of Helsinki in 1836.[22] After several years in Uppsala Törnegren returned to Finland, then traveled in 1841 to Moscow in order to study Russian. Upon his return to Finland he got a job in the university library and began to do research which was for the first time in Finland explicitly literary-critical. His dissertation for the docentship in the history of learning, the 77-page *Primordia artis scenicae Hispanorum* (1843), dealt with the birth of the Spanish theater and constituted the first Finnish study of modern literature. In his analysis of the development of Spanish and its influences from Arabic, Törnegren ascribes Spanish rhymed verse also to Arabic influence. Törnegren names Juan Ruiz the first Spanish playwright; Ruiz's coarse vernacular *Danza general* is the only surviving medieval Spanish play. Among later dramatists Törnegren pays particular attention to Bartolomé de Torres Naharron, who becomes a kind of touchstone in the dissertation for comparison with other writers. Jerónimo Bermúdez, Juan de la Cueva, and Cervantes (with his historical drama *La Numancia*) are also discussed in detail; Lope de Vega is all but dismissed in a passing reference.

The 68-page dissertation Törnegren wrote for his associate professorship in the history of learning, *De initiis carminis epici apud Italos* (1845), deals with the early stages of Italian epic verse. Törnegren notes that since the feudal system was not as highly developed in Italy as it was in France, the epic, with its ideological reliance on the institution of chivalry, never attained in Italy the national significance it had in France. Therefore, the roots of early Italian epic must often be sought further north: *Bovo d'Antona*, for example, is only a variation of the English romance *Bevis of Hampton*, and both Charlemagne and Roland play significant parts in Italian epic verse. Törnegren identifies the great Italian innovator as Boccaccio, whose life and work he depicts with reference to Baldell's study. He then analyzes in detail the *Orlando innamorata* of Boiardo and the *Orlando furioso* of Ariosto before concluding his history of the early stages of "national" epic with the Christian *Gerusalemme liberata* of Torquato Tasso. Törnegren's focus throughout is on nationality; and his final conclusion is that at the end of the Middle Ages and the beginning of the modern era there simply was no great Italian "national" epic of the type and caliber of Camões' *Os Lusíadas*.

When Blomqvist died and Törnegren was appointed acting professor of the history of learning, he published two more monographs in Swedish. The first of these, a history of troubadour poetry, *Trubadurerne* ("Troubadours," 1849), was never finished; the second, a study of the early development of Swedish-language poetry in Finland (1849), was extremely brief. *Trubadurerne* was published in a 62-page booklet and broke off in mid-sentence just as Törnegren was beginning to discuss the troubadours and their poetry. In his introduction to the booklet he notes the increasing interest European nations are showing in studying the "lost days of their infancy," and how this interest has led quite rightly to a renewed consideration of Provencal and the rich poetic tradition written in it – that "first-born daughter of modern European poetry" (2). Törnegren promises to describe the birth of Provencal and its early literary use, the flowering of Provencal poetry in the twelfth century, and the characteristic features of troubadour poetry. He was also planning to translate samples of troubadour poetry into Swedish. In his truncated work he only really deals with the first of these, the birth of Provencal, although, in the course of discussing the conditions under which troubadour poetry arose he does also touch on the institution of chivalry.

The study is by and large a compilation of French scholarship on Provencal culture. Törnegren does in passing draw a comparison with Finland: the fact that, like the Finns with Swedish, the Basques and Celts had successfully resisted the incursion of foreign and more powerful languages showed, Törnegren claimed, that "the languages of the nations contain the seed of imperishability" (9). His central sources include the works of Raynourd and above all Fauriel's *Histoire de la poésie provencale*; the latter had appeared in Paris only three years before Törnegren published his study, and its influence over Törnegren's conception of his subject increases the closer he gets to his main topic, troubadour poetry. Despite its truncation at a critical point, Törnegren's study is interesting both in its choice of topic and in its plan of execution. The topic had been the focus of considerable study in Germany and France, but despite its historical and geographical distance, it offered striking possibilities for current local insight, particularly into the problematic of the endangered native language. The subject was later picked up by greater Finnish critics, first among them C. G. Estlander less than two decades after Törnegren.

The 18-page article he published in 1849 entitled "Teckningar af den Svenska vitterhetens äldre öden i Finland intill Franzén" ("Notes on the History of Swedish Literature in Finland Until Franzén") is, despite its brevity, the first history of Finnish literature in which the bibliographical approach used by earlier historians of literature was displaced by an explicitly historicizing one. Törnegren favored literary-historical topics in his teaching, and had begun to gather data for a larger history of Swedish-language Finnish literature. Upon the appearance of his article, the Turku publisher J. W. Lillja asked Törnegren to write a more comprehensive study of the same subject, and it was apparently in response to this request that Törnegren undertook further research. Of especial importance was the lecture course he gave in the 1851-1852 academic year on the history of Swedish and Swedish-Finnish literature – the first course ever offered on this topic in Finland.

When the professorship in the history of learning was closed in 1853, Törnegren continued in the head librarian post that was split off it until his death in 1860. In the language wars that raged throughout this period he was a Svetophile, and for that reason was portrayed in several contemporary Fennophile accounts (that of S. G. Elmgren, for example) in a rather negative light. His language politics led him into a bitter debate with the influential nationalist philosopher Johan Vilhelm Snellman; according to Elmgren he lacked all patriotic fervor and all desire to work for the betterment of the fatherland. Loosely paraphrasing A. I Arwidsson's famous pronouncement –

Fredrik Wilhelm Pipping (left), and Carl Wilhelm Törnegren (right), two professors of the history of learning. During their tenure, and that of professor Alexander Blomqvist, the discipline shifted in focus from the history of books and science toward the history of literature. (M)

"We are not Swedes; we do not wish to become Russians; therefore let us be Finns" – Elmgren suggests rather snidely that, if he could not be a Swede, Törnegren would rather be a Russian than a Finn.[23] When in the late 1850s Törnegren began his lecture course on Swedish literature, which was cut short by illness, he predicted with some bitterness – and perhaps with some ironic exaggeration – that in a few generations the Finnish cultural elite would no longer be able to read its own Swedish classics without a dictionary.[24]

The work of historians of learning bore fruit for emerging literary criticism in bibliographical form as well. The most significant of the bibliographies prepared in this period was Pipping's Finnish-language index of books printed in Finnish by the mid-eighteenth century, *Luettelo Suomeksi präntätyistä kirjoista kuin myös muutamista muista teoksista, joissa löytyy joku kirjoitus Suomen kielellä, tahi joku johdatus sitä tuntemaan* ("Index of Books Printed in Finland, and a Few Other Works in Which is to be Found Some Text in the Finnish Language, or Some Reference to Such," 1856-57). Elmgren was in a position to watch Pipping complete his life's work, but was initially skeptical of its value. In the fall of 1854 he characterized Pipping's chronological method of listing books primitive. When the work was completed, however, Elmgren changed his mind: the Finns, he now said, possess a better list of their own books than any other nation (*Notes* 124, 269).

The earliest bibliography of Finnish literature is probably the Swedish Johannes Schefferus's *Svecia literata* (1680), a list of all Swedish authors and works arranged by king, but its index included a section entitled "Writers of the Finnish Nation." Schefferus lists about forty Finnish writers, selected on inconsistent principles and presented with inadequate bibliographical information. Another significant aspect of Schefferus's work is that when Johannes

Mollerus reprinted it in 1698 in Hamburg, that list of forty Finnish writers established the existence of a specifically *Finnish* literature outside Scandinavia.

Another important early bibliography is *Aboa literata* (1719) by the Swedish historian Anders Anton von Stiernman. Its very title (Aboa) points to the dominant role played by the University of Turku in Finnish publications of the seventeenth and eighteenth centuries. Von Stiernman's bibliography was the first separate list of works published in Finland, and throughout the eighteenth century it served as the central reference work on books published in the seventeenth century.

These early bibliographies are especially important today due to the destruction of vast amounts of the oldest Finnish literature in the 1827 Turku fire, which destroyed the entire university library. Of the roughly 40,000 volumes then stored in the library, only the 379 works in 827 volumes that were out on loan survived.[25]

Literature and Nationalism

In Finland the status of literature and literary culture was complicated throughout much of the nineteenth century by the country's bilingualism: the uneducated majority spoke Finnish, the educated elite Swedish. This rendered the entire question of a "national literature" in conjunction with a "national state" – an issue of central social importance in the debates of the 1840s – highly problematic. Part of this problem was internal: how Finns, separated from their old mother country Sweden in 1809 and now an autonomous grand duchy of Russia, were to arrange their linguistic, cultural, and political affairs in an internally coherent fashion – how to mediate, for example, between the Swedish-language cultural tradition of the past, which should not be lost, and the emergent Finnish-language political (and increasingly cultural) realities of the present, which should not be stifled. Part of it was also external, international: how to interact with the old mother country, and with the new one?

Until the 1830s Finnish literary life was conducted on a relatively modest scale. Finnish literary texts and readers were few. The country's population crossed the one million mark in 1815 and the million and a half mark in 1845. Due to the active work of the Lutheran Church, literacy was relatively common, but specifically literary literacy was still rare. According to one calculation, in the mid-nineteenth century of the 1.6 million Finns only about 3000 belonged to the cultural elite who read books (Öller 1-25). Around the same time the professor of history and writer Zachris Topelius reached slightly higher figures: scholarly literature, he estimated, might reach 400 book-buyers and 1000 readers, popular nonfiction and newspapers a total of 1400 buyers and 14,000 readers, and literary texts 3000 buyers and 20,000 readers (Öller 1-25).

But the national and historical imperatives proclaimed by European romanticism were already beginning to give literature a new social role to play. Nearly a hundred years earlier, in his *Fragmente über die neuere deutsche Literatur* (1767), Herder had portrayed literature as the reflection of a nation's unique spirit and stage of historical development; and in 1817 the Finnish journalist and historian A. I. Arwidsson (1791-1858) insisted in his dissertation on medieval romances, which was heavily influenced by Schelling, Friedrich von Schlegel, Geijer, and Molbech – *Ingenii romantici, aevo medio orti, expositio historica* – that earlier times and deeds determine in precise detail the nature of every phenomenon.

Thus literature began to be seen as the *reflection* of the nation, but also, more actively, as the indispensable tool for *shaping* the nation. This latter more active conception proved crucial in Finland's sociopolitical situation in the first half of the nineteenth century: split off from one country and attached to another.

One interesting perspective on this issue of the role literature was to play in the period in question was provided in Johan Jacob Tengström's (1789-1858) lengthy article in the Aura calendar for 1817-1818, "Om några Hinder för Finlands Litteratur och Cultur" ("On Some Obstacles for Finnish Literature and Culture"). Tengström was a philosopher and a historian who was appointed docent at the University of Turku in 1811 and professor of philosophy in 1827; his familiarity with the philosophy of Hegel was later to prove influential for Snellman.[26] In his essay on the "hindrances" or obstacles to Finnish literature and culture, Tengström sought a synthesis of classical ideals and contemporary culture. He makes frequent appeal to Greek classics, but simultaneously raises the issue of nationality, insisting that Finnish culture has been and will always be shaped by its environment – by its lakes and rivers, hills and mountains, fields and forests. "For the character of a nation, like that of any organic formation, is determined as much by the nature that surrounds it as by its primordial, innate, 'seminal' or 'root' tendency."[27]

Tengström remained fairly pessimistic about Finland's chances at generating its own literature. He believed that the Finnish reading public would always be too small in number to maintain its own independent literature, and Finns would have to turn to other nations, where the "writer's profession pays better." Despite this pessimism, however, Tengström outlined for Finnish literature a series of tasks that were to inform the cultural-political debate of the next few decades. He argued that it would be advantageous for the nation's religious, ethical, and social development for its knowledge to be increased by education and its imagination to be stimulated by art, particularly through the publication of the old folk poetry:

> So far no Finnish writer or citizen has seriously contemplated compiling and publishing a more comprehensive collection of Finnish poems, though such a collection would be as pleasing as possible not only to every Finn but to the art critic, the historian, the philosopher, and the linguist. The prompt execution of such a project is essential, before the knowledge that there once was a national poetry in Finland be reduced to the status of ancient fable.

In his remarks on the development of Finnish culture written twenty years later and appended to a history of Finnish academics, Tengström articulates a far clearer sense of literature's central significance in shaping a national identity. The first edition of the *Kalevala* had been published only three years before Tengström wrote, in 1835, and he proclaimed that now the poetry of

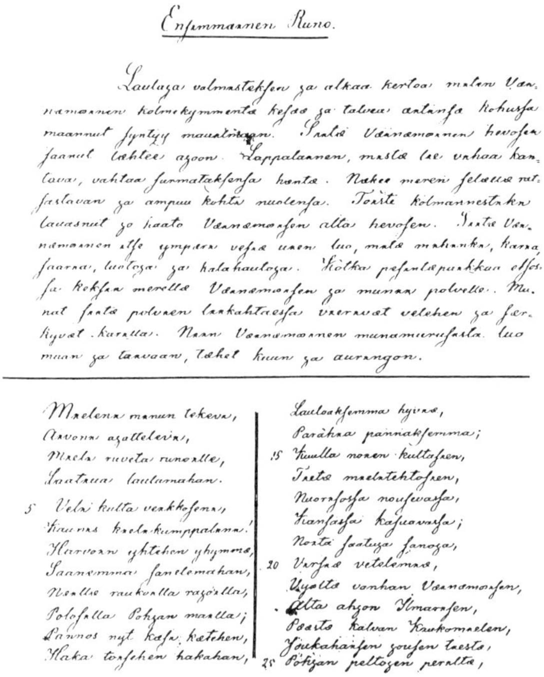

Elias Lönnrot and his manuscript of the *Kalevala.* (SKS)

the Finnish people was "in its first flush of strength, while yet only a seed or bud in its immediate natural form and figure." In the fragmentary bits and pieces of the old poems the Finnish people were now seeking reinforcement "for the significant story of their forefathers through their own unique civilization."

The focal debate for the development of a Finnish literary culture in the early nineteenth century revolved around the concept of a national literature, and the results of this debate had a significant effect on the formation of nineteenth-century literary criticism as well. The concept of literature itself was different then from now. What was called in Swedish "litteratur" and in Finnish, once the term was coined, in the 1831 founding of the Finnish Literature Society, "kirjallisuus," was considerably broader in scope than what we today call literature, and did not distinguish *belles lettres* from other writing.[28] Tengström, for example, argued that the most effective way to begin creating a national literature for Finland would be through books on history and geography; and the Finnish Literature Society, which aimed at "achieving a national literature," stated as its primary purpose the dissemination of information about Finland and its history.[29]

Tengström was not the only early nineteenth-century writer to ponder the requirements for the creation of a national literature and to stress the importance of literature for the development of culture. There was widespread agreement with him as to the weak material foundation for Finnish literature, but there was also a pervading sense of urgency. Finnish literature was needed, and it was needed fast. Johan Gabriel Linsén (1785-1848), associate professor and later full professor of Roman literature, believed fervently in the late 1810s that a group of romantic lyric poets might indeed soon appear in Finland. Around the same time, in 1819, Gustaf Renvall (1781-1841), a student of Finnish and associate professor in education, stressed that a national culture could only be built on the foundation of an emergent contemporary literature. Renvall's studies of Finnish hymnody for the 1836 hymnal revision committee – especially "Försök till Finsk Prosodie" ("Essay on Finnish Prosody," 1819) and *Välmenta råd för Finska Psalm-Författare* ("Well-Meant Advice for Finnish Psalmists," 1830) – were far and away the most significant formal analyses of Finnish poetry in the early decades of the nineteenth century.[30] The first to speak of a Finnish national literature in *Finnish* was professor of history Gabriel Rein (1800-1867), in his article "Om National-Literaturens i Finland utveckling, och nödvändigheten af dess vidare bearbetning" ("On the Development of a National Literature in Finland, and Requirements for its Further Cultivation," 1842). Rein admitted the difficulty of injecting a complex cultural heritage into the undeveloped Finnish language, but argued strongly for its necessity.

The most vocal proponent of Finnish literary nationalism was the philosopher and statesman Johan Vilhelm Snellman (1806-1881). Snellman's remarks on a Finnish national literature are scattered through the essays he wrote from the 1830s to the 1870s, but the most significant in terms of the contemporary debate, especially that over the formation of a literary-critical discipline at the university, are the essays of the 1840s, which had a powerful effect on the intelligentsia of the day. Snellman's sovereign impact on the cultural-political realities of the day has perhaps been exaggerated: he argued

Johan Vilhelm Snellman (M)

strenuously, for example, for the creation of a native Finnish literature, and the chair that was founded was aimed from its inception toward aesthetics and world literature. (To be sure, the first appointee, Fredrik Cygnaeus, published important analyses of Finnish literature as well.) On the other hand, Snellman's importance for the development of Finnish literary criticism should not be underestimated, either; and it will pay us to look more closely at his contribution to the birth of the Finnish critical institution.

In his article for *Saima* Magazine in 1844, for example, "Det enda nödvändiga icke blott för den inhemska litteraturen" ("The Sole Necessities Not Only for Finnish Literature"), Snellman includes in the concept of national literature everything that "expresses the characteristic spiritual development of of a nation." Like most writers of the day, Snellman did not draw a clear boundary between *belles lettres* and nonfiction. He insists that religious awareness, legal concepts, conceptions of world and national history, and poetic feeling and thought develop differently in different cultures, and that the literature arising out of each area (theology, law, history, poetry) will therefore reflect a nation's spiritual life. Two years later, in an 1846 article that appeared in the *Kallavesi* supplement to *Saima,* "Inhemsk litteratur" ("Finnish ['Domestic'] Literature"), Snellman begins by asserting that a people's spiritual energy arises out of its awareness of its own national character. This awareness takes shape in two ways. The first is in relation to other peoples, as an experience of difference and independence. This is, however, in some sense a negative path to national self-awareness, based on rejection of the alien. The second is through literature:

> In literature, on the other hand, a nation learns to know its own thoughts, its own particular ways of conceiving the nature of beings,

> learns to understand its own customs, its own institutions, its entire existence, history. Now it appears that only a conscious striving toward this rational self-awareness might coax out of the people all the inward powers of the spirit, and history shows that it is precisely when a nation has thus expressed the being of its own spirit that it can most effectively work for the best of human development and take its place among the civilized nations of the earth.

The literature Snellman is referring to here is not just poetry – not just folk poetry, for example, which from our present point of view was then the greatest contribution to date to Finnish national literature – but all writing that expressed the current stage of Finnish development. Snellman insisted in the 1844 piece that centuries-old classical literature was inadequate to express a nation's contemporary spirit. The idea of a Finnish national literature originated, Snellman wrote in the 1846 essay, in the late eighteenth century, when the "gust of nationalist sentiment" then beginning to blow in Europe had excited interest in Finland as well in the Finnish language and Finnish geography and history. Still, the situation by the mid-nineteenth century was alarming: the native Swedish-language literature was capable only in rare exceptions of promoting the nation's efforts at self-determination, and Finnish-language literature was nonexistent and giving no signs of springing into life. His burst of collective self-criticism in the 1844 essay is indicative:

> It is, therefore, idle to look further [than the Swedish speech of the cultural elite] to explain the low level of literature in our country. We write too little and read too little; we are slow, sluggish; we devote all our energies to furthering our careers; book sales are pathetic, book publishing not worth mentioning, royalties nonexistent – what is all this but the result of these two causes: Finnish culture is underdeveloped, we lack a national literature, and these two go hand in hand.

There is some disagreement as to the influence of Hegel on Snellman's philosophy and politics of language; some scholars have argued that Herder and Fichte were at least as important in the formation of Snellman's thought as Hegel. The most recent research, however, Lindman's in 1981 for example, points to Hegel as the crucial influence; and certainly Snellman's conception of the development of culture and the existence and force of a national literature is dialectical in something like the Hegelian sense.

Snellman placed national literature and world literature in the same dialectical relationship. National literature is born in international interrelation, in an interactive recognition of one's own expression, and conversely a well-developed national literature bears fruit across national boundaries.

Snellman's theory and politics of literature were instrumental in the historical development that has consistently assigned literature a significant historical and social role to play. Since Snellman insisted that the Western cultural heritage and Finnish peasant traditions must meet and merge in literature, that literature mediate between high and low, the cultural elite and the people, his insistence on making literature the medium of social debate and development also had the immensely significant effect of giving literary realism pride of place.[31]

The Wasenius Bookstore was the first bookstore in Helsinki. It was founded in 1823, four years after the first bookstore was opened in Turku.

Since the international base for Finnish literary culture was connected with Sweden, it was natural that this connection continue after 1809. The Swedish writers Geijer, Almqvist, and Tegnér were extremely influential in Finnish literary life in the first half of the nineteenth century, and extensive analyses were written of their works in the Finnish press. A stay in Sweden was crucial to Snellman's intellectual development, and the germ for antisveticist Fennomania seems itself to have come from Sweden, above all from Geijer.[32] Sweden was also important as cultural and intellectual mediator: new developments in the literary and scholarly life of Central Europe typically came to Finland through Sweden. In Finland as in Sweden, Romanticism was often tied up with classical ideals:[33] of the great pioneers of Finnish literary culture, Johan Gabriel Linsén was a professor of Roman literature, Axel Gabriel Sjöström was a scholar and later professor of classical languages, and J. J. Tenström and J. L. Runeberg were trained in classical philology.[34]

The university's connections with Sweden also continued. Professors had long moved freely between Turku and Swedish universities, and even when the university was moved to Helsinki in 1828, partly in order to move the center of Finnish intellectual life farther from Stockholm and closer to St. Petersburg, scholarly contacts with Finland's western neighbor remained strong.[35]

A look at the borrowing records from regional students' club libraries from 1831 to 1852 shows the importance for Finns of the period not only of Swedish but of German culture as well. In Palosuo's tabulation, the list of most-borrowed books looks like this:

Finnish borrowings		Foreign borrowings	
Runeberg	92	Goethe	147
Kalevala (S)	35	Almqvist	145
Kanteletar	30	Schiller	129
Kalevala (F)	29	Fr. Bremer	115
Franzén	26	Shakespeare	49
Gottlund	23	Heine	40
Snellman	18	Hugo	31
Juteini	15	Tegnér	25
Fr. Cygnaeus	13	Oehlenschläger	23

(S) = in Swedish (F) = in Finnish

German and Swedish literature clearly dominate. Since the late Middle Ages German universities had provided a channel of advanced (graduate-level) education for Finnish scholars, and particularly in the late eighteenth century the cultural interchange between Germany and Sweden/Finland was brisk.[36] During the autonomous period as a grand duchy of Russia these cultural connections were never severed; in fact, the Academy of Science in St. Petersburg simply opened up a new channel between Finland and German scholarship.[37] The popularity of German literature in Finland is evident in many less public ways as well. The language scholar Anders Johan Sjögren (1794-1855) read Herder in the 1810s and immediately became an ardent admirer of German literature, in so doing giving a significant push to the study of folk poetry among the Turku Romantics.[38] Beginning in the 1820s Goethe and Schiller were especially popular among the Finnish cultural elite, and perhaps Schiller more than Goethe, a preference that had much to do with the cultural-political situation. As the scholar Rafael Koskimies wrote in 1958, Schiller was more congenial than Goethe to "the spiritual frame and needs of emergent nationalism"; he offered "watchwords to the young, newly awakened national consciousness" (100).

As Magnus Björkenheim's study of *Äldre fransk litteratur på herrgårdar i Finland* ("Older French Literature in Finnish Manor Houses," 1929) shows, a knowledge of French culture was an integral part of the Finnish cultural elite's education, particularly beginning in the eighteenth century. These traditions remained strong throughout the nineteenth century as well, and for an emergent Finnish literary criticism French literature too offered an attractive subject.

In addition to classical antiquity, then, Swedish, German, and French literature formed the foundation on which nineteenth-century Finnish literary criticism was built, first in the form of newspaper articles, then of scholarly research. Needless to say, however, critical interest extended beyond these countries as well. English literature and philosophy had found their way to Finland in the mid-eighteenth century, though the metaphysical Kant seemed to interest Finns more than the empirical Locke.[39] One of the earliest scholarly pilgrims to England and ardent admirer of English literature in Finland was F. M. Franzén. Finnish Shakespeare-criticism was initiated in a 44-page study of Hamlet in the Joukahainen calendar for 1845 by R. I. Holsti (1815-1852), a talented young aesthetician and philosopher who died at the age of thirty-seven as rector of a Hämeenlinna school.

As the nationalist movement gained ground in Finland, debate arose as to the extent to which foreign literary culture was needed alongside a native tradition. This question was particularly pointed in conjunction with the issue of translation – the significance of translated foreign works for the development of a Finnish literature. While the value of good translations for the development of the Finnish language and the broadening of Finnish thought, dissenting voices were heard as well. Pietari Hannikainen, a journalist and playwright who endorsed J. V. Snellman's strictures, wrote in *Kanava* magazine in 1847:

> If in fact the entire Swedish literature were translated into Finnish, what would we have gained thereby? Not much more than we have by reading it for six or seven hundred years. Only a literature that arises from the people themselves can speak to all their needs, their desires, their hopes and fears. It is that kind of literature that we had better start working for, for that is the only kind of literature that would deserve the name *Finnish* literature.

It is a commonplace of historical scholarship that the Russian government looked favorably upon the nationalist movement in Finland, because it had the effect of distancing Finland from Sweden. Before long it also became clear that Russian culture also offered Finns new possibilities, even new contacts with the cultural centers of Central Europe. The Academy of Science in St. Petersburg became a significant channel for the top flight of Finnish scholars; particularly the disciplines of history, folklore, and linguistics benefitted from the open eastern border.[40]

It is important to recall, as Henrik Grönroos makes clear (74-75), that Sweden-Finland had shown intellectual interest in Russia at least since the seventeenth century. Jouko Hautala has documented Porthan's plan in the late eighteenth century to undertake a comparison of Finnish and Russian folk poems (69). Beginning in the 1810s contact with the Academy of Science in St. Petersburg opened up new opportunities for Finnish scholars, and the first (relatively modest) surveys of Finnish literature to be undertaken in the Russian period were published in St. Petersburg: Anders Johan Hipping's *O finskoj literature* ("Finnish Literature," 1820) and A. J. Sjögren's *Ueber die finnische Sprache und Ihre Literatur* ("On the Finnish Language and its Literature," 1821).

Nevertheless, as scholarly and other contacts with Russia increased in number through the nineteenth century, it began to seem to some commentators that Finnish culture's relations with Sweden and Russia constituted, and implicated individual scholars in, a kind of political statement. Zachris Topelius, for example, asserted in the 1840s that Svetophilia was Finland's spiritual armor against Russia.[41] Indeed around the middle of the century there was a counterreaction to the antisveticism of the nationalists, and a large segment of the young intelligentsia realigned themselves westward, toward Sweden and the free-thinking debate on science and politics then under way in Central Europe.[42] By the end of the nineteenth century these various inputs and alignments had led to a strikingly new cultural climate.

In the genesis of the Finnish literary-critical institution described above, the press's role became increasingly important. Literary and periodical publications were released in ever greater numbers for the use of emergent social groups, and the press played an important part in this development. Newspapers and magazines published poems and stories as well as rudimentary "literary notices" that by the middle of the century had evolved into book advertising on the one hand and reviews on the other. By the mid-nineteenth century journalistic book reviews were already quite competent; at their best they were lengthy essays on trends in current literature, and (then as now) they provided literary critics with one of their most significant forums. By the 1880s the press had established itself as an integral part of the literary-critical institution, and with its highly visible public debates it regulated literary ideals and values.

The First Professorship

In 1828 Swedish academic circles began to discuss the establishment of professorships in aesthetics and modern literature at the universities of Uppsala and Lund; and in 1835 such a position was indeed established in Uppsala, with the poet P. D. A. Atterbom the first appointee. The chair's official title was "moderna litteraturen och aestetik" (modern literature and aesthetics).[43] A similar proposal was made in Finland in 1836, patently influenced by the Swedish example, but also more generally by the enhanced status of art brought about by romanticism, and simple university politics.

In the international scholarly community literary history was rapidly becoming a field of central importance, and the critical focus was shifting from classical antiquity ever closer to the contemporary period. Above all, in this period literature was fast becoming an immensely significant social phenomenon whose nature and structure cried out for careful study. The 1830s were a particularly important period of expansion for Finnish literature: the Finnish Literature Society was founded in 1831, Lönnrot's *Kalevala* was published in 1835, and J. L. Runeberg's poems began to appear.

In January 1836 professor of medicine Immanuel Ilmoni proposed the creation of three new professorships to the university senate: in botany, pharmacology, and aesthetics.[44] The first two had been proposed the previous year; the aesthetics professorship was a new proposal. Ilmoni was familiar with the Swedish debate on the creation of a similar post, and informed the senate of an article published by the first Swedish appointee, Atterbom, in the course of that debate entitled "Om rätta betydelsen af vhad man kallar ästhetik, betraktadt såsom föremål för akademisk undervisning" ("On the Correct Meaning of What is Called Aesthetics, Considered as a Subject of Academic Study," 1831). In it Atterbom argued for the establishment of a discipline broad enough to encompass the philosophy of beauty and art, the history of "humanistic" literature in general, the history of poetry in particular, the history of the other arts, mythology, symbolism, and the history of the Swedish language and literature.[45] Concurring with this point, Ilmoni proposed that the new post's field be conceived broadly as aesthetics.

In his statement to the university senate Ilmoni argued eloquently and at some length for the importance of the new post. The "philosophy and history" of various art forms was, he said, not merely excellent, but eminently useful, in the attention it paid to the origins of humanity in, and its development out of, the mysterious era of myths into the bright day of modernity. Familiarity with the field was essential to "the education of every cultured person," and of especial importance to clergymen, who should be versed in works like Milton's *Paradise Lost* and Klopstock's *Messiah*. One suggestion Ilmoni made for funding the new post was to cut theology professors' salaries back to the level of other professors – an impolitic move, for Ilmoni's proposal promptly met with stout opposition from the theologians.

In the spring of 1836 the university senate forwarded Ilmoni's proposal directly to the chancellor without making its own recommendation on it; the senate promised only to give a statement if required. This rather unusual administrative procedure was evidently motivated by the senate's unwillingness to establish new professorships. The proposal came before the senate again in December of the same year, but it was more than a year before it was even introduced into discussion, and by that time, on March 10, 1838, Ilmoni's opponents were ready for him. The main counterargument was presented in a 36-page statement by A. A. Laurell, professor of theological dogmatics. A professorship of aesthetics, he said, was entirely unnecessary; the individual's aesthetic awareness was independent both of personal aspiration and of external instruction, and in any case this sort of instruction was better suited to an academy of art than to the university. The existing professorship of the history of learning was more than adequate, Laurell wrote. In conclusion he asserted that the number of professorships had already reached an upper limit that could not be exceeded without negative consequences.

In the discussion that ensued not a single professor defended Ilmoni's proposal. The most common position was that further thought should be given to the matter of what new posts were most important. A few professors' arguments are worth special mention. J. J. Nordström, professor of national economy, agreed that a professorship in aesthetics was needed, but that it should not be top priority. Professor Johan Gabriel Linsén, the first director of the Finnish Literature Society, insisted in his statement on the importance of the field of aesthetics – art had always had a significant impact on the development of human culture, and it was especially important for young people – but voted against the formation of a new professorship in the field. He defended his negative vote by appealing to Atterbom's 1831 article: the history of Swedish literature could well be included in the purview of the Swedish post, but no such literature as yet existed in Finland. It was certainly possible that at some future date a greater Finnish literature would be born, but at present the local cultural prerequisites for such a post were not yet extant.

Gunnar Castrén suggests that Ilmoni's agitation for the new post may have been partly motivated by the fact that he had a candidate in mind, the promising young poet Johan Ludvig Runeberg (1804-1877), who was then 32. Runeberg had begun his studies in Turku and in Helsinki had completed two excellent dissertations on classical literature, *Specimen Medeam tragoediam Euripidis cum Medea Senecae comparans* (1830) and *Specimen academicum observationses quasdam circa Chorum tragicum continens* (1833), but had been passed over – unfairly, apparently – for an associate professorship in classical philology.[46] In the late 1830s, during the debate over an aesthetics professorship, Runeberg had just been appointed lecturer in Latin in Porvoo, and his letters show that he would have preferred to stay in Helsinki in a university post. But the university senate's position was clear. When the proposal finally came to a vote in the fall of 1839, only Ilmoni voted for the creation of a professorship in aesthetics, two professors voted for the creation of a professorship in botany, and twelve voted against the proposal *in toto.*

We know very little about the second phase in the history of debate on the aesthetics chair. In the mid-1840s J. J. Nervander, professor of physics and himself a poet, seems to have proposed the creation of a chair in literature, but his proposal is not to be found in the university archives.[47] A private statement was requested from Fredrik Cygnaeus, docent in history and an eloquent advocate of national culture; in his statement he compares classical and modern literature and comes down strongly in favor of the latter. Circumstances, he wrote, were now utterly changed from the day when Latin was the common language of the civilized world; nowadays "scarcely anyone would voluntarily choose Latin as the verbal vehicle of his living heart." In this situation students must be given the opportunity to drink "from the bubbling fountain of modern literature."[48]

The third and critical phase in the creation of the aesthetics chair began in late 1848, when the university senate took up professor Laurell's proposal. Laurell argued that the funding for a new post could be found by cutting back the practical instruction in German, English, and French. This sort of cutback would make possible "more scholarly instruction in modern literature" than had thitherto been possible with the "diletanttish" literary knowledge of practical language teachers. The new post would provide for a combination of "the aesthetician's sensitivity to beauty and the broad erudition of the historian."[49] Laurell's brief covers all the ground that was eventually to be incorporated into the new discipline: aesthetics, the history of literature, and philological literary scholarship.

While the senate was considering the new post, in January of 1849, the important Swedish-language morning newspaper *Morgonbladet* printed Fredrik Berndtson's editorial calling strongly for a literature professorship and reflecting on the discipline to be associated with it. According to Berndtson, "literature" (*den sköna litteraturen*) had long been neglected at the highest academic levels in Finland – despite all the "rich and wonderful" opportunities it offered for the advancement of culture and knowledge. A chair in "aesthetics and the history of that and the arts [*de sköna konsterna*]" was therefore essential. In addition to literature, Berndtson said, scholarly instruction was needed in the visual and plastic arts and music. Art is an indispensable vehicle of progress: without an awareness of and sensitivity to art the student's "conception of the world, of life, and of history" invariably remains superficial and distorted. Berndtson was opposed to the inclusion of the modern languages within the purview of the post: if the post were created at the expense of lectureships in the modern languages, language learning would inevitably suffer. He also had his candidate for the job: "if a professorship in aesthetics were created, the university might have the honor of bringing within its sphere the country's most excellent literary personage, our noblest poet, J. L. Runeberg, who would raise the institution's glory and prestige in

Johan Ludvig Runeberg (SLF)

no small amount . . ."

The senate now generally recognized the need for such a post. The main obstacle was funding. Laurell's suggestion in regard to the cutting back of language teaching was not well received; another suggestion, however, put forward by the university rector, professor of history Gabriel Rein, was met with enthusiasm. His plan was to discontinue the professorship of the history of learning and in its place create two posts, a head librarianship with salary and status commensurate with professorial rank, and the professorship of literature and aesthetics. Rein argued that the history of learning had become such a broad and inclusive field that it was no longer possible for one person to master it.[50] It would be more productive to have every professor teach the history of learning in his own field.

Only one professor opposed the creation of this new post in the senate debate, law professor Wilhelm Gabriel Lagus. It was impossible to understand the modern world, Lagus argued, unless one was well versed in classical antiquity; and in any case aesthetics and modern literature were an odd combination, since aesthetics was a branch of philosophy.

When the new post was created in 1852, its title was modeled not on the Uppsala chair but on the form more common in Central Europe: its field was specified as "estetik och nyare litteratur," aesthetics and modern literature, or, in Finnish, "estetiikka ja nykyiskansain kirjallisuus" (aesthetics and the literature of the modern peoples).

When the first appointee, Fredrik Cygnaeus, began work in 1854, the inaugural invitation written by professor of Greek literature Nils Abraham Gyldén made it clear that special emphasis in the new post was to be placed on aesthetics. The history of literature had already been a part of the old history of learning, but aesthetics, with its profound influence on human culture, was now for the first time represented at the highest academic level in Finland.

PART TWO
The First Generation of Literary Critics

Fredrik Cygnaeus

Fredrik Cygnaeus, the first holder of the chair in aesthetics and modern literature, was born in 1807 in Hämeenlinna, a vicar's son.[51] In 1820 his father was appointed bishop of the Evangelical-Lutheran congregations in St. Petersburg, and so the young Fredrik came to spend his adolescence in a cosmopolitan city. He was conscientiously educated at home, with considerable attention paid to the modern languages. In 1823 Cygnaeus entered the University of Turku, where he followed the Swedish and German debate over romanticism with great interest. After the Turku fire Cygnaeus continued his studies in history, philology, and philosophy at the new University of Helsinki, both in class and in the unofficial but prestigious "Saturday Society," where the leading young Finnish literati met and discussed the important issues of the day. In the summer of 1831 he moved to Sweden to pursue further studies, and was planning to continue on from there to England to study English, but returned at the end of the year to Helsinki, where he took his Master's degree in 1832. For the next five years Cygnaeus taught history, geography, and statistics at a cadet school in the small Eastern-Finnish town of Hamina. There he wrote his first literary study, on the first great verse epics of Runeberg – though his graduate studies still lay ahead of him, and were until the late 1840s to be in history, not in aesthetics or literature.

Having failed to find a teacher's post in the Helsinki area, Cygnaeus resigned his job at the cadet school and moved to Helsinki as an independent scholar, where he worked on his docent dissertation on Hannibal, *Commentationis de Hannibale pars prior indolem ejus a scriptoribus infamatam vindicatura* (1839). His declared intention in the work was to rehabilitate Hannibal's reputation from the slanders of Roman historians. Only one-fourth of the 132-page dissertation deals with Hannibal's career; the rest focuses on the classical historians who wrote about him, especially Livy and Tacitus. When the dissertation had been approved in the fall of 1839, Cygnaeus was appointed docent in history at the University of Helsinki. Around the same time he was also appointed rector of the Helsinki trivium school, which provided his primary livelihood.

In the early 1840s Cygnaeus continued his studies in history. On a lengthy research trip he made to Central and Southern Europe about this time, the archival work he did in Paris, the Vatican, and Montpelier, which was intended to elucidate the imperial history of Sweden, ended up being overshadowed by literary work and his self-familiarization with Central and Southern European art.

When professor of history Gabriel Rein was appointed rector of the university in 1848, Cygnaeus was chosen as his deputy, and served as acting professor of history for the next four years.

The early 1850s were a productive period for Cygnaeus. He published poems, a long but unperformable historical closet drama (*Claes Flemings tider,* 1851), and two volumes of popular essays, the second of which (1853) contained the significant piece on the *Kalevala,* "Det tragiska elementet i Kalevala" ("The Tragic Element in the *Kalevala*").

Cygnaeus defended the dissertation he wrote for the newly created professorship in aesthetics and modern literature, *Erik XIV såsom dramatisk karakter* ("Erik XIV as Dramatic Character"), in 1853. His official faculty opponent was Carl Wilhelm Törnegren, who praised the dissertation in a brief statement to the Division of History and Linguistics, calling the topic so felicitous and the treatment so excellent that along with Cygnaeus's other merits it undisputably qualified him for the professorship in aesthetics and modern literature.[52] In the senate discussions of the appointment, as in the senate report to the chancellor, Cygnaeus's earlier work was considered inadequate qualification for the post, but his excellent professorial dissertation, along with a few articles, constituted more than adequate proof of his qualifications for the post.[53]

Particularly in the early years, Cygnaeus was a popular teacher. In his notes for 1854, Elmgren reports that Cygnaeus's lectures on modern literature (contemporary German literature, at that point, with emphasis on the works' aesthetic value and significance) were attracting more students than his earlier lectures in history.

Cygnaeus remained fairly prolific a writer in several areas throughout his university career. In addition to his literary works, he published historical studies, literary criticism, and book reviews, the most significant of which had scholarly substance to them as well. At some point, however, the administrative and pedagogical burden of the job began to tire him. In 1860, at age 53, he applied for either leave of absence or early retirement, and was on leave all through 1861, in the spring term of 1862, and in the year of his retirement, 1867. Beginning in the late 1850s Cygnaeus served as dean of the Division of History and Linguistics, and in 1865-1866 he was vice rector of the university. After his retirement in 1867 he continued to write and publish, and as chairman of the Art Society campaigned for an arts academy in Helsinki – to no avail. He died in Helsinki in 1881.

In his 126-page professorial dissertation Cygnaeus set himself the task of probing the suitability of the Swedish King Erik XIV for dramatic treatment; his texts were four plays revolving around that king. In introducing his analytical method he says he will concern himself with literary minutiae only insofar as is absolutely necessary; his main concern is with the broad principles that structure the plays he is examining (43).

The dissertation is written in a flowery but leisurely style, with many digressions: in the middle of an analysis of a play, for example, Cygnaeus will go off on a polemical tangent to warn against too stringent an interpretation of Aristotelian poetics. But in fact, compared with the baroque rhetoric of some of his articles, the dissertation is almost spartan in its simplicity and clarity.

The dissertation shows Cygnaeus's thorough familiarity with German literary scholarship and aesthetics. It is peppered with citations from Hegel, A. W. Schlegel, Hettner, and Gervinus. On the other hand, Cygnaeus is no slavish imitator of other people's work. Starting from Hegel's dictum on complexity as the key to a character's vitality – "eine solche Vielseitigkeit allein giebt dem Charakter das Intresse der Lebendigkeit," only so multifaceted a portrayal charges a character with the feeling of life – Cygnaeus asks what makes several Swedish kings (not only Erik, but Gustaf Vasa and Gustaf III as well) their patent suitability as dramatic figures. His suggestion is that while larger-than-life personalities are generally on the wane in world history, Sweden has continued to produce them in the modern era. Examining how Erik's persona has been "reflected through various poetic minds," Cygnaeus describes the four dramatic Erik-constructs as "idealizations in the free play of poetry" (5-6). He stresses the constant interchange of opposites in Erik's inner and outer life: Erik was "so weak and yet so strong." Cygnaeus reads the four plays as masterful analyses of the inner life of great personages – a life full of "warmth, vitality, interest, strength."

Cygnaeus's historical studies had clearly shaped his approach to the Erik plays. He insists that literature in general and *belles lettres* in particular cannot be isolated from its historical context. His historical method is not, however, positivistic, or overly concerned with the minutiae of historical event; rather, it is solidly grounded in the philosophy of history. He will ask, for example, how a historical transition manifests itself in great personages. This is especially clear in his comparisons between the Italian playwright Alfieri's *Saul* and the Erik plays. Cygnaeus sees Saul, like Erik, as a watershed figure between two periods, and as a result both have a "fragmented inner core" with broken aspects of both periods (40). This notion is so important that Cygnaeus returns to it at the end of his dissertation; Schiller, he says, succeeded brilliantly in his plays in executing his theoretical principle according to which "noble souls" emerge triumphant even in the moment of collapse (123).

Through his four exemplary plays Cygnaeus works to illustrate four basic human types common to drama. Of the four, one, by the Swedish playwright Bengt Lidner, is written in the spirit of French classicism; the other three, by Prutz, Börjesson, and Beskow, are romantic. Cygnaeus's own standpoint is closer to the romantic than the classical. In his lengthy discussion of Lidner's *Sorgspel*, he challenges the classical notion that a theoretical framework – specifically, Aristotle's *Poetics* – can prescribe rules for artistic endeavor. For Cygnaeus Aristotle is compelling only in bits and pieces, for the purpose of heuristic mental gymnastics, not as an ironclad set of rules. He blames Lidner's failings, particularly his nonexistent "sense of history," on precisely this slavish adherence to Aristotle (69-83). The other plays he finds alive with the "warm inner" light of romanticism (Börjesson). He praises Börjesson, and all play-

Fredrik Cygnaeus (M)

wrights beginning with Schiller, for portraying the people in a positive light. Here he cannot forgo a reference to Finland and Runeberg: "To Finland belongs that high human honor, that one of her poets has dared to sing, 'äfven med tiggarn är Gud' [God walks even with beggars]" (97). Cygnaeus criticizes Prutz for projecting on the king in *Erich der Bauernkönig* only his own ideals of freedom. Beskow's *Erik XIV*, on the other hand, is to Cygnaeus's mind Sweden's finest historical drama, for in it the author has successfully blended "the real demands of history with the ideal demands of poetry" (120-122).

One of the most striking claims Cygnaeus makes in the dissertation is that Calderón may have based his play *La vida es sueno* on Erik. He grounds his conjecture partly on the fact that, in the original Spanish, the events of Calderón's play are set in distant Moscow, not, as in the contemporary Swedish translation, in Castile.

The professorial dissertation was Cygnaeus's only monograph on non-Finnish literature. Several of his articles and journalistic pieces, however, were to have a significant impact on the history of Finnish literary criticism. Most notable among these were his analyses of Runeberg's *Älgskyttarne* ("Elk Hunters") and *Hanna* (1837) and *Nadeschda* (1841), the lengthy article on the tragic element in the *Kalevala* (1853), and numerous reviews, the best known of which was the unfinished piece introducing Aleksis Kivi's *Nummisuutarit* ("Heath Cobblers," 1865). This last was significant not only as a literary-political defense of an early Finnish realist but also as an excellent analysis of Kivi's work. In general, in fact, Cygnaeus showed such an unusual perspicacity and sensitivity in his readings of contemporary literature that he has been called an early pioneer of close reading (von Platen 156).

The importance of the essays on Runeberg's verse epics, on the other hand, was that they were the first extensive and carefully argued analyses of Finnish literature, and that they were written in a kind of historical degree zero: Runeberg was the first significant Finnish writer and had just begun his career, and the critical hierarchy of literary value was still unformed.

The earlier of these two essays, the slightly over 60-page "Ett och annat angående Älgskyttarne och Hanna" ("This and That About the 'Elk Hunters' and 'Hanna'"), appeared in Cygnaeus's first book, the literary album *Jääkynttilät, ströskrift i fria häften* ("Ice Candles, Free-Bound Pamphlets," 1837). The later, "Om Nadeschda," appeared in two issues of the *Helsingfors Morgonblad* in early 1841. In both articles Cygnaeus lauds Runeberg as a great, internationally significant writer whose greatness is grounded in an inner unity between his work and the nation. In Runeberg's work the Finnish-speaking people of Finland have received an accurate, authentic, and profoundly perceptive portrait. He chastises the Finnish intelligentsia for not recognizing one of their own, and the Swedish press for not recognizing the living poetic spirit of the Finnish-speaking people (along with nature, of course) as the source of Runeberg's poetic genius. Cygnaeus places particular emphasis on the complexity of Runeberg's literary portraits: that he does not merely idealize the people, does not paint a purely positive picture of them, but shows their weaknesses and failings as well, often with a light ironic touch. Here Cygnaeus seems to be anticipating his later critical position that literature is most powerfully driven by an unresolved conflict.

In defending Runeberg's work Cygnaeus decisively rejects the notion frequently found in Nordic reviews that Runeberg is imitating Goethe, Voss, or Scott. Insisting that only the *Odyssey* could offer an obvious model for the *Elk Hunters*, he charted a specifically Finnish path for Runeberg criticism – one that was to deviate strikingly from its Swedish counterpart until the early twentieth century.

The cultural-political significance of the earlier Runeberg piece was further enhanced by the close connections Cygnaeus drew between Runeberg's work and contemporary society. The small educated elite still had only the vaguest conception of the people on and around which the new nation was to be based, and according to Cygnaeus Runeberg was now offering the first reliable image:

> Since untold centuries have beaten their icy wings fleeing across it [Finland] – since, in a word (even leaving out of consideration that political insignificance to which Finland has always been, and will always be, condemned), the noble, strong-bodied, poetic generations of Finnish men have long since vanished into the farthest reaches of time, and what is left is wrapped in the pale shroud of tradition – so must the *Kalevala* and the *Elk Hunters* become the Herculaneum, the Pompeii, which those who long to gaze into the living eyes of a long-buried nation must bring into the warm light of day. (107)

This national, cultural-political polemic is central to the *Nadeschda* pieces as well. Cygnaeus pays a great deal of attention to the formal interplay of dramatic and epic elements in the work, but still the fulcrum of his argument is cultural politics:

> That he [Runeberg] should have developed in this way is of no small cultural-historical significance to Finland. For the nation's budding sense that the world of intellect is its true vital element requires a great deal of support on which it can rest with trust and joy. Without that support our nation's aspiration toward elevated intellect will dissipate into hollow amusement. (May 24, 1841)

In 1849 Elias Lönnrot published his revised, expanded, and final edition of the *Kalevala*, which gave the Kullervo motif a much more comprehensive and coherent expression than in the earlier edition. Shortly after, Cygnaeus began work on a study of the figure of Kullervo in the epic in order to test its applicability to the romantic axiom that the Finnish poetic *Volksgeist* was definitively epical in character. The 60-page article "Det tragiska elementet i Kalevala" appeared in 1853, in the second volume of Cygnaeus's essay collection *Afhandlingar i populära ämnen* ("Disquisitions on Popular Themes"). According to Cygnaeus's first biographer, Emil Nervander, this article constituted Cygnaeus's "breakthrough in Finnish literature" (285). One striking feature of the article was of course the fact that it dealt with a literary work written in Finnish; but even more important was Cygnaeus's integration of current national tendencies into his grandiloquent analysis. In his review of the essay collection in the *Morgonbladet*, August Schauman writes that "we willingly forgive the author if we fail to understand every sentence, or to see through every image, for reading his words is so abundantly satisfying to both the head and the heart."

In the earliest-written section of the article, a brief discussion of the "instability of certain theories" that was written shortly after the publication of the revised *Kalevala*, Cygnaeus brings his irony to bear on the notion that literary theory should prescribe for literature. One of the established received truths in Finland had long been the purely epical nature of the Finnish literary genius: Finns had no inclination to lyrical or dramatic creation. Cygnaeus proclaims that "the sword that slashes the sorely tried strings of this misbegotten son of Kalervo's heart must slash also that whole warp and woof of theory woven as the restraining wrap of the Finnish people's poetic spirit" (78).

In his analysis of Kullervo Cygnaeus draws heavily on Hegel's dialectical conception of history. Like Hegel, Cygnaeus understood conflict to be the driving force of history. In various contexts he stressed that "in the annals of history struggle means life," or that the history of Christianity was "the mortal struggle of the doomed to immortal victory over death," or that the peaceful period of Swedish history in the late eighteenth century had raised a crop of "moral weeds."[54] Much the same vision informs his study of Kullervo. Cygnaeus argues that Kullervo's tragedy lies in his inner division: nature made him a hero but fate made him a slave. He was born to act and wield his "natural giant strength," but his lowly birth as a slave makes this innate virtue a vice. From beginning to end he aspires to do great deeds, but has neither family nor country for which to perform them. All through his discussion Cygnaeus stresses the creative unresolved conflict between plot-advancing elements that characterizes the entire story of Kullervo. It is, he says, fortunate indeed that the dusty theories of tragedy propounded by the schoolmen had had no shaping power over the creation and performance of the *Kalevala* songs, since the "natural poets'" instinct had grasped with sure confidence how the tale must proceed by unresolved conflict.

As Cygnaeus notes, this is altogether different from Zachris Topelius's interpretation in the *Helsingfors Tidningar*, which saw in the Kullervo story a prechristian sermon on the importance of a good home, in the spirit of 1 Corinthians 13. Without mentioning Topelius's name, Cygnaeus sets himself in staunch opposition to this sort of historical anachronism and "saccharine sweetness." In Cygnaeus's vision happiness is gained through sacrifice and struggle, the highest happiness and fulfillment through mortal risk, the braving of death. He argues that the imposing figure of Kullervo could only have been created by a people who had achieved through wars and suffering an "unshakable joyous courage." In his interpretation Kullervo is elevated into a representative figure for the entire Finnish people, for Finnish sufferings and sacrifices, out of which proceeds their ultimate happiness.

> By a rich life we by no means intend the life of a rich man, though on the other hand we do not second that all too common view that constant hunger is the best incitement to the poetic life and death by starvation its noblest end. Even so, the life we mean finds its greatest riches and strength in precisely that which often destroys physical well-being and health: storms, struggles, great setbacks. (72)

Cygnaeus returns to this theme in his conclusion, where the idea of intertwined victory and defeat is once again extended to apply to the romantically conceived "author" of the Kullervo story, the Finnish people:

> And while the dust of the hearts and heads of those through whom the tragic element of the *Kalevala* burst forth is now as nameless as it is mute, still this element will have an enlivening and elevating effect so long as a single Finnish heart still beats and a single Finnish head still strongly grasps great ideas. (131)

Despite his stated belief that he is reading the story of Kullervo like nature, without mediation, Cygnaeus's analytical method in the article is interestingly text-based. At the end of his study he says that he does not intend to compare the Kullervo story with other tragedies he knows, since that would be like comparing nature herself with certain of her individual products. Such a comparison would inevitably prove unfavorable to the latter. But alongside this idealized romantic view, Cygnaeus had an astonishing sensitivity to the interrelations of the poem's narrative elements, and his analysis of the artistic structure of the piece remains persuasive today. He notes, for example, that the poetic narrator – or, as he says simply, the "poem" – takes Kullervo's side no matter how vicious his deeds (and how detailed the narrator's description of those deeds), but also how the narrator's voice is often infected with the negative opinion of the other characters in the story. Though Swedish was his first language, Cygnaeus seems to have felt at home in the poem's Finnish. He bases his ironic refutations of earlier commentators' claims about the conflicted Kullervo plot on detailed semantic analyses of the poem's Finnish diction, and also on a comparison of the original with two translations, Borg's Swedish and Schiefner's German. He also declares that the undeniably contradictory nature of certain plot elements in the Kullervo story is immaterial when apprehended with the "aesthetic faculty."

Cygnaeus published his study of Runeberg's *The Songs of Ensign Stal, Om "Fänrik Ståls Sägner,"* in 1861, the year after the second series of Runeberg's poems about the War of Finland had appeared. Back in 1848, when the first series appeared, Cygnaeus had planned to review it; now his review became a 161-page monograph. It was the first study of what was to become in Finland Runeberg's best-known poetic work, and of Cygnaeus's literary-critical works it was the clearest, most consistent, and most coherently structured.

Cygnaeus reads Runeberg in three ways, which correspond to the three parts of his study: "general," "closer," and "detailed." The third, his detailed reading of individual poems, is in turn divided into three sections, based on class distinctions among the central characters in given poems: "low-born," "middle-class," and "high-born." In this respect the focal poem in Cygnaeus's reading becomes "Fänrikens marknadsminne" ("The Ensign's Market Memory"), in which representatives of all three groups – a poor war invalid, the ensign-narrator, and a general – are brought together. Cygnaeus sees the poem as a "mediating bond" connecting the varied elements out of which Runeberg creates his poem (359). Cygnaeus presents another argument for reading the collections as a unified whole: there is a poem in the second series, "Landshöfdingen" ("The County Governor"), which brings both civil and military activity under the large rubric of patriotic action – whereas in the first series these two were treated as two distinct activities.

Cygnaeus is not afraid to make bold, even harsh evaluative judgments of the poems. He praises the poems' patriotic fervor, their love of country, to the skies; but attacks strongly everything that does not fit easily into his nationalist-idealist interpretive framework, such as excessive sympathy for the enemy. The overt patriotism of Cygnaeus's study is unscholarly, certainly, in a narrow sense of that word; it could be explained away, if perhaps rather uneasily, by noting that he was aiming his words at a general rather than a specialized professional readership. A more important point to make about Cygnaeus's reading, however, is that in stressing the subjective protest, the fruitfulness of the conflict between the individual and society, out of which Runeberg's poems (like the Kullervo story in the previous work) arise, Cygnaeus is opening a powerful new dimension in Finnish literary criticism. Literature for Cygnaeus is no aestheticist idyll into which one retreats from the clashing world around one. He sees the societal conflicts that are built into literature and literature as playing a vital role in those conflicts – and stresses the positive value of this interaction.[55]

When, late in his professorial career and in the early years of his retirement, Cygnaeus undertook the study of Aleksis Kivi's (1834-1872) work, he found himself in much the same situation in regard to Kivi as he had been thirty years earlier in regard to Runeberg. Kivi's first published work, the 1864 *Heath Cobblers*, was the first Finnish-language literary work to rise above the mediocrity of the desultory literature written in Finnish in the period. But in addition to having a tiny Finnish-language readership, Kivi had in August

Colonel von Döbeln in the Battle of Juutas. Albert Edelfelt's illustration for Runeberg's *Tales of Ensign Stal*, 1890s.

Ahlqvist, professor of Finnish and himself a poet, a powerful opponent – one who, out of a pedantic fetish with "correct" (i.e., decorous, dignified, educated) grammar and style and an idealized conception of art, also, perhaps, out of simple professional envy, never tired of calumniating Kivi's work. In this situation Cygnaeus's long, analytical, and sympathetic reading of *Heath Cobblers* (1865) – like his later introduction to the second (1873) edition of *Seitsemän veljestä* ("The Seven Brothers"), first published in 1870 – had an immense and salutary impact. Largely as a result of Cygnaeus's piece, Kivi was awarded a prestigious prize for the play (beating out the now venerated Runeberg, the expected winner) which was to constitute the major part of his meager income in the productive decade of his short life. Cygnaeus's article remains one of the best, most perceptive, most profound pieces of Kivi criticism written to date; he uncovers Kivi's roots in European literature, but also stresses the powerful folk realism that elevated Kivi far above his precursors. It is significant to note that, outside of Cygnaeus's work on Kivi, there was very little understanding of literary realism in Finland in the 1860s and 70s.

In his professorial capacity Cygnaeus familiarized himself with a broad spectrum of modern literary history, and lectured in both of the fields specified in his job description, aesthetics and modern world literature. Between 1854 and 1866 he lectured on German literature, French literature, eighteenth-century literature with special emphasis on the best dramatic works, the aesthetics of drama and lyric, and the aesthetics of Vischer and Hegel. He lectured on Vischer for the first time in 1855, and on Hegel in the 1856-1857 academic year. Throughout the rest of his teaching career either Hegel or Vischer provided the theoretical foundation on which he based his discussions of aesthetics. In addition to those two, his most important theoretical mentors in aesthetic questions were Herder, Schelling, and Schiller, and the Swedish poet-philosophers Geijer and Tegnér.

This may be the place to note that beginning in the 1850s Vischer was the leading aesthetician for Finnish literary critics; his massive work *Aesthetik oder Wissenschaft des Schönen* (1847-1857), which applied Hegelian philosophy to aesthetics, was studied closely. No doubt the popularity of Hegelianism in Finland contributed to the scholarly attention this book received. Vischer's classicism, his emphasis on harmony, may also have helped predispose Finnish critics against Kivi's realism, and generally against the realistic movement in the later decades of the nineteenth century.

While Cygnaeus paid homage to Hegel as a pioneer in aesthetic theory, his attitude toward Hegel was fundamentally a critical one. The substance of his critique of Hegel is probably most clearly outlined in the "lectio praecursoria" he gave at Oscar Toppelius's 1856 dissertation defense, published that same year in the *Litteraturblad* (148-155). Cygnaeus deplores the increasing formalism of literary criticism, which he ascribes to Hegel's influence: because of Hegel "persons with no deeper poetic sensitivity . . . had appointed themselves judges in the poetic world of expressions of creative genius that remain foreign to their narrow conceptions." Cygnaeus argues that, since Hegel was born and raised in an era whose primary task was "to smash the universal hegemony of French neo-classicism," it is not surprising that he could bring a fine sensitivity and understanding to the art of classical antiquity and yet respond negatively and polemically to the romantic world view and its literary products. The main problem, Cygnaeus said, was not Hegel's political conservatism; his antipathy toward the Schlegels and their ideological kin reflects "the distaste of a dry Protestant mind for the free play of Catholically inclined romanticism."

In his own readings of and theoretical pronouncements on literature Cygnaeus untiringly attacked excessive intellection and systematization, and defended in their stead creative individualism, emotion and imagination. For him literary interpretations should be intuitive, arising out of and based firmly on insight. The interpreter of literature should also strive to understand as far as possible the author's intention. In the study of *The Songs of Ensign Stal* Cygnaeus takes the following position on normative, prescriptive criticism:

Aleksis Kivi, who founded modern Finnish literature in the 1860s almost without predecessor or peer. (SKS)

One wonders when the arrogant criticism of today will at last deign to surrender itself up to the object it studies, to let itself be caught up in the spirit of the works, to appreciate them from the inside. It seems to seek and find its whole satisfaction from the enjoyment of its own superiority and the demonstration with absolute crystal clarity that the author of the work has scarcely exhibited a schoolboy's knowledge of the correct way to proceed in the construction of a more or less flawless work. Only he, the critic, has come into possession of this secret, with regard not only to the work at hand, but to all those that already exist or that may someday exist. Nor is he at great pains to put this lamp of his under a bushel. It must shine in its full glory upon all humanity in order to show two things: first, and most important, that he the judge understands all; second, that the reviewed author has more than anyone else failed to grasp a smidgen of what he should have done. Little or no attempt is made to sympathetically discover what gives a given product its characteristic color, dignity, and life. The only concern is how the product should have been manufactured in order for it to have any value whatsoever. It is assumed in advance that the person who has allowed the blossom of his entire being to bloom in the work

> dare not mutter against that incomparable wisdom that is gained by absent-mindedly browsing through the work for a moment or two. (29-30)

In Cygnaeus's aesthetic conception, art is a higher and nobler reality to understand which requires authentic enthusiasm and inspiration. It is, he believed, the artist's right and duty to idealize reality, to improve it in his art. This conception is also evident in his approach to historical drama: historiography must depict reality as it is, art as it should be.

Later estimates of Cygnaeus have typically found him too honey-tongued, too idealistic, too much a man of the grand vista, to be called a literary scholar in the strict sense of the word. He was also, for his time, unusually concerned with contemporary literature: Runeberg, the *Kalevala*, Kivi. Still, Cygnaeus's position as the founder of Finnish literary criticism is incontrovertible. In that "lectio praecursoria" for the Toppelius dissertation defense he defends the status of his field as follows:

> It seems to me that literary history is one of those branches of scholarship that might be practiced in our country with at least as much success as in any other country in the world, and perhaps with greater success than in some.
>
> Considering our isolated geographical location and the difficult circumstances in which we Finns live, the same cannot be said about a very great number of branches of human knowledge. (148)

In Finland it should, he said, be easier to write literary history than great literature, which requires for its creation a spirit elevated by "great circumstance" and an experience charged by "massive conflicts." The necessary preconditions for literary history, on the other hand, were already in place: "Those features that characterize a nation should already, or soon, be sufficient to generate writers in the field of literary history who will stand up in comparison with the best other countries have to offer" (149).

Toppelius's dissertation was the first written under Cygnaeus's direction, and these introductory remarks to it may well be read as a kind of declaration of independence for Finnish literary criticism.

Critics as Faculty

During Cygnaeus's professorial term, and to some extent before it, a group of scholars in the new discipline began to win their academic spurs. The first docent of aesthetics, appointed in 1847, was Fredrik Berndtson; in modern literature, Theodor Clasen was appointed the first docent in 1852. These dates suggest that the need for the new discipline was clearly recognized even before the professorship was created. In 1853, the year before Cygnaeus was permanently appointed to the professorship of aesthetics and modern literature, Berndtson deputized in the post for the first half of the year, Clasen for the second half.

Fredrik Berndtson (1820-1881) was born in Falun, Sweden, and upon graduating from high school in Uppsala in 1840 moved with his parents to Finland, where he matriculated into the University of Helsinki.[56] He com-

Fredrik Cygnaeus speaking at the student celebrations, the Flora Day, May 13, 1848, the occasion of the first singing of the Finnish national anthem, "Our Land," with lyrics by J. L. Runeberg. Drawing by A. Federley in the 1880s.

pleted his Master's degree in 1844 and took a job with Sven Gabriel Elmgren as reporter on the *Helsingfors Morgonblad* (moving in 1845 to the *Morgonbladet*) and worked as an assistant librarian at the university library. Particularly in the Helsinki cultural life in the 1840s Berndtson was a central figure; besides being a scholar and a journalist, he played an active part in a regional students' organization and wrote poems and plays. After his publication in 1847 of two dissertations on aesthetics, clearly influenced by Hegel – *Adumbratio notionis phantasiae pulchrae* and *De Vi et Essentia ingenii in arte se manifestantis* – Berndtson was appointed docent in aesthetics. On the *Morgonbladet* Berndtson made ample room for articles on aesthetic, literary-theoretical (e.g., "Den moderna romanens karakter" ["The Character of the Modern Novel"] 1849, "Några ord om familjelitteraturen i allmänhet och i Sverige" ["Some Words on Family Literature in General and in Sweden"] 1852), and literary-historical issues (e.g., "Balzac och den Fransyska Romanen" ["Balzac and the French Novel"] 1846, "Frans Michael Franzén" 1847, "Johan Jacob Nervander" 1848, "Percy Bysshe Shelley" 1848, "Jacob Frese" 1849, "William Makepeace Thackeray och hans skrifter" ["and his writings"] 1850, "C. J. Almqvist" 1851), to the point that the paper began to get a literary reputation. In the winter of 1850 Berndtson gave an open public lecture on the nature and existence of the human soul, in which he sought with the aid of "the organizing power of science" to bring order to that intellectual chaos that he claimed the new literature had sown in people.[57] As a scholar Berndtson was more of an aesthetician than a literary historian; one of his few writings on literary history to appear outside the *Morgonbladet* was a 23-page study of Franzén (1849).

In the early 1850s Berndtson was Fredrik Cygnaeus's only rival for the professorship in aesthetics and modern literature. Upon submitting his application for the post to the university senate in December of 1852 Berndtson informed the senate of his intention of writing the required professorial dissertation during the spring term of 1853.[58] Berndtson had recently accepted the position of editor-in-chief of the official government paper *Finlands Allmänna Tidning*, however, and failed to write the dissertation by the end of the respite period; he was disqualified for the position.

In the mid-1850s Berndtson was appointed department head in the Office of the Governor General, an appointment that grew out of his editorship of *Finlands Allmänna Tidning* (the two jobs were combined at this time and the incumbent of the latter position automatically assumed the duties of the former). Berndtson's work kept him busy until 1865, but he still found time to translate A. Kuhn's aesthetic treatise on the idea of the beautiful into Swedish (1865). In his late work *Dramatiska studier och kritiker* ("Dramatic Studies and Reviews," 1879), he reprinted his reviews of dramatic performances on the Helsinki Swedish stage over a 35-year period; this work remains a fascinating source on the early history of the Finnish theater.

Theodor Clasen (1827-1855) was born in Oulu,[59] a small town in northern Finland that for its size produced an astonishing amount of new blood for Finnish cultural life in the nineteenth century. Clasen graduated from high school at age sixteen and in the mid-1840s came to Helsinki to study history and classical philology; these studies he supplemented with private tutoring in the modern languages. He completed his Master's degree at age twenty.

The first indication of Clasen's scholarly interests came in the fall of 1850, when he gave a talk at the Ostrobothnian students' club on the origins of the romantic epic. The talk traced the history of French medieval literature, including the literature of Provence. At the same time he was working on a dissertation on English medieval literature: *De Galfredo Chaucero poeta statuque ante eum poesis anglicae* (1851). The completed dissertation shows Clasen's thorough familiarity with the early stages of English literature and the English secondary literature in the field, and he also deals with the influence of French and Italian medieval literature on English literature of the period. According to his faculty opponent, Törnegren, for a first-book writer Clasen possessed a remarkable talent for presenting his thoughts clearly and fluently.

Beginning in the early 1850s Clasen had worked at various jobs, as an assistant librarian in the university library, as a reporter for the *Morgonbladet*, and as an English teacher at a Helsinki lyceum. He worked for the *Morgonbladet* for the entire year of 1852, and during that time he published in its pages both short stories and literary studies; the latter reflected his two main interests, Anglo-Saxon literature and folklore. Another testimony to Clasen's interest in folklore is his fellow student's reminiscence that he knew Herder's book *Stimmen der Völker in Liedern* almost by heart. In 1852 the pieces he wrote for the *Morgonbladet* included an article on the "romantic fairy-tale play," based around two Shakespeare plays; and the next year he wrote an article for the *Litteraturblad* on Robert Burns, inspiring students with sample poems published along with the article and a simultaneous lecture course on Burns's poetry.

Clasen's interest in English literature was not an isolated case. In the mid-nineteenth century there were a good number of Finnish students and readers of Anglo-Saxon texts. Runeberg had wondered in an 1842 article whether Macbeth was a Christian tragedy; in 1845 R. I. Holsti had published a lengthy article on Shakespeare; and beginning in the 1830s the *Helsingfors Morgonblad* had been introducing English poets to the reading public, in an overview article in 1833, Byron in 1834 and 1836, Shakespeare in 1835, and Scott in 1836. In the early 1850s the *Litteraturblad* introduced not only Burns, but Cooper (1851), Thackeray (1854), and Dickens (1855). Dickens had first been discussed in the Finnish press back in 1840, in *Vanadis* magazine.

In November of 1853 Clasen was appointed Berndtson's successor as acting professor of aesthetics and modern literature. We have no reliable information on his lectures – in the university class schedule he announces that he will be lecturing on "topics requested by the audience" – but we do know that he lectured at least on English literature from its beginnings to the eighteenth century. According to Clasen's student, Wilhelm Bolin, who was later to become a professor of philosophy, Clasen's lectures differed from the standard practice of the day in being based on carefully prepared notes but still combining clarity with vivacity.

In the spring of 1853 Clasen had traveled to England to do research. The following summer he traveled to the United States in the hope that a warmer climate would cure his worsening tuberculosis. But not even a stay in Cuba in early 1855 had an effect on the progress of the disease, and Clasen died in the U.S. in the fall of 1855. He was a promising scholar whose illness prevented him from showing how far his talents could take him.

The first scholar to defend his doctoral dissertation in the new discipline was Oskar August Toppelius (1828-1904). Toppelius was the son of a municipal doctor from Oulu, and had moved to Helsinki in 1843 to study the history of learning and modern languages, taking his Master's degree in 1850. Toppelius was interested in the theater – in the 1860s he drew up a plan for the creation and curriculum of a bilingual theater school in Helsinki, and when in 1866 the school opened its doors, with numerous problems partly due to the lack of Finnish-speaking students, Toppelius served as its director. He also argued strongly for the translation of Shakespeare's plays into Finnish for the use of the Finnish-language stage.[60] After a research trip to Denmark and Germany in 1851-1852, he started work on his dissertation on Ludvig Holberg.

The 67-page study that resulted, *Ludvig Holbergs komedier* (1856), is an overview of Holberg's comedies. Toppelius's central sources are the earlier studies of Holberg by Prutz, Rahbek, and Werlauff, and Heiberg's history of literature. Toppelius's analyses of individual plays is, however, remarkably original work. After reviewing Holberg's life and dramatic performances of his plays, Toppelius focuses on literary influences on and characterization in the plays themselves. He draws connections between Holberg's drama and the Italian folk theater, Molière, and Plautus, while insisting on the Danish plays' national character. This insistence clearly reflects the emphasis placed by Cygnaeus, who directed the dissertation, and by the era in general, on the national character of literary texts. Though Holberg drew heavily on themes and motifs from earlier dramatic literature, his plays nevertheless have "a characteristically Danish way of presenting things" (33); they have raised "the people's literary self-esteem" (63); in reflecting "the unique quality of his own people" Holberg had become the father of Danish literature (1).

Oscar Toppelius (M)

In his analysis of Holberg's characterization, Toppelius argues that Holberg was a master of comedy but not a humorist in the positive sense of the word: not conciliatory, forgiving, loving. His comedy does not expand into a coherent world view; he failed to balance the weaknesses and shortcomings of his characters against some sense of human ideals and lofty aspirations. Toppelius's idealistic concept of humor seems to reflect a characteristic Finnish attitude, and it was soon to plant a seed in fertile soil: Aleksis Kivi studied literature under Cygnaeus and Toppelius in the late 1850s, read Holberg avidly, and charged his own work with precisely the kind of complex affirmation of human follies and goodness that Toppelius missed in Holberg.

Toppelius's docent dissertation, the 70-page *Folkvisorna om Marsk Stig* ("Ballads about Sheriff Stig," 1861), also dealt with Danish literature. At the same time it ranged widely through a field of particular interest to scholars during Cygnaeus's professorship: the medieval folk poetry of the European nations. Toppelius is concerned with the published versions of the folk ballads about Sheriff Stig from Vedel's in the sixteenth century to Grundtvig's in the mid-nineteenth: their medieval (thirteenth-century) historical background, relations among the various editions, and the history of the ballads' development (Toppelius suggests some revisions in Grundtvig's recent reconstruction). He also examines the ballads' image of historical reality and their characterizations. Toppelius concludes his dissertation with a look at the use of the Sheriff Stig motif by later Danish authors. In an appendix he reprints three textual variants of the ballad.

Toppelius brings a theoretical interest as well to his major concern, the ballads' relation to historical reality. Like most mid-nineteenth-century Finnish theoreticians who had been reading Hegel, Toppelius places great emphasis on the relationship between the collective and the individual, the public and the private. Historiography registers great events and subordinates little events to them, whereas "poetry seeks out small groups and individual figures and – despite the fact that the public and politically significant is just as familiar to it as anything else that is human and great – in its portraits of life and events it goes deeper into the individual than historiography and presents to us whatever of the intellectual [*idéeuppfyldt*] and beautiful it finds there" (15). In its alignment with this field of interests between the collective and the personal Sheriff Stig's fate is tragic: he finds himself forced to defend his personal honor and love, but thereby also forced to relinquish his loyalty to the king; he defends the sanctity of home and hearth, and this puts him at cross purposes with the state and fatherland. Toppelius's interpretive model is familiar from Snellman and Cygnaeus: in it tragedy arises out of a conflict in which revenge is simultaneously both a duty and a crime.

Starting in 1861 Toppelius served as docent of aesthetics and modern literature; in the same year he was appointed assistant librarian in the university library. In 1867 he published a lectorial dissertation on Homer, *Achilleus i nionde sången af Iliaden* ("Achilles in the Ninth Song of the *Iliad*"), which was to be his last critical study of literature. Toppelius begins with the established critical claim that the ninth song of the *Iliad* is spurious and inappropriate to the epic as a whole, seeking through philological and narratological analysis, as well as polemics against earlier critiques, especially that offered by Georg Grote, to show that the ninth song does indeed intregrally and organically fit into the epic. His position is that the ninth song foreshadows Achilles' punishment for the reader, and therefore is not extraneous at all but a well-motivated part of the whole.

Toppelius wrote sketchy notices on the content of his teaching for the university class schedule. In the 1861-1862 academic year he promised to lecture on whatever the audience desired, and he continued this practice until 1864-1865, when he provided the slightly more informative notice that he would be lecturing on "modifications of the beautiful." In lecture courses held later in the 1860s he taught the "factual subjects" (*kunskapsämnen*) of the discipline, Swedish drama of the Gustaf III era, and Vischer's aesthetic system. This last was his teaching program for the 1868-1869 academic year, which was his last at the university. Even these cursory notices indicate that, like Cygnaeus, Toppelius divided his time pedagogically between the two disciplinary branches: aesthetics and modern literature.

Upon approval of his lectorial dissertation at the end of the decade, Toppelius left Helsinki to take up a lectureship in Greek at the Vaasa high school, a post he would work in until he retired. He wrote no more literary criticism.

Johan Viktor Calamnius (1838-1891), who defended his doctoral dissertation toward the end of Cygnaeus's professorship, was born the son of the vicar of Ii.[61] Calamnius's scholarly inclination was toward aesthetics. In his dissertation, *Försök till framställning af Aristoteles' lära om det sköna och konsten* ("Inquiry into the Presentation of Aristotle's Theory of the Beautiful and Art," 1865), he painstakingly summarizes all of Aristotle's definitions of the beautiful, beginning with the general (conceptions of beauty in general and of art in particular, conceptions of the nature and purpose of art) and concluding with the specific (the various art forms). Theoretically the dissertation rests most heavily on the authority of the German thinkers Edward Müller and Franz Biese. Calamnius's method is primarily descriptive. He does attempt to estimate Aristotle's applicability to the present, but also stresses that the important task is one of explaining Aristotle objectively and reliably. "In this sort of presentation the educated reader who is familiar with modern aesthetics can best find Aristotle's true meaning" (v).

The dissertation might well be read as a prolegomenon to Calamnius's pioneering work as a translator. His Finnish translation of Aristotle's *Poetics* (1871) was blasted by August Ahlqvist, who charged that translating Aristotle into Finnish in the cultural situation of the early 1870s was like gentrifying a "farmhouse in the forest," which lacked hundreds of more essential objects, with a piece of "ancient Greek statuary."[62] Still, Calamnius had the right idea: he sought to generate for Finnish aesthetic studies a whole cluster of Finnish terminology. Unfortunately for Calamnius's reputation, later generations rejected almost all of his suggested terms.[63]

All through the 1860s Calamnius played an active role in many sectors of university life and Helsinki cultural life. He translated Dierckx's Belgian work on the problems caused by bilingualism into Finnish (1863), was one of the founders of and early reports for the *Kirjallinen Kuukauslehti* (Literary Monthly, 1866-1880), and from 1868-1874 was president of the Ostrobothnian students' club. Even after exchanging his academic career for an ecclesiastical one – he was appointed notary clerk for the Kuopio cathedral chapter in 1875, and eventually became Dean of Kuopio – he continued to work with the subjects that had interested him at the university, translating Ovid and Plato into Finnish.

Critics as Library Staff

Even after the university's head librarianship and the professorship in the history of learning had been officially split into two posts, research on and related to the history of literature continued to be carried out by library staff. Törnegren may have been instrumental in passing on this tradition, though personally he published no more literary scholarship; in any case, the tradition was continued by Sven Gabriel Elmgren, Gustaf Edvard Ingelius, and Karl Collan.

Elmgren (1817-1897) was the son of an associate pastor from Parainen.[64] Parainen is in a purely Swedish-speaking area of Finland, and Elmgren never did learn to speak Finnish, though he himself found his lack of Finnish bothersome and, inspired by Snellman's Finnish nationalism in the 1840s, tried to learn it. Elmgren wanted to become a minister and took his Master of Divinity degree, but decided against the ministry and for several years after graduation worked simultaneously on Berndtson's *Morgonbladet* as a repor-

Degree congregations were the ceremonial high points of university life. Above, the congregation committee for 1864, including literary critics A. A. Laurell (back row, first on left), J. V. Calamnius (back row, second from right) and J. J. F. Perander (front row, first on right) (M). Below, A. Federley's drawing of the ceremony at the Student Union, 1880s.

ter and in the university library as an assistant and later associate librarian.[65] Having strenuously expressed his Fennomania in various journalistic pieces, he was asked by Snellman to be a reporter for the central Finnish cultural paper, *Litteraturblad*, which he did from 1850 to 1854. During Elmgren's period on the paper it focused almost entirely on literature and neglected political and social issues; this was partly due also to intensified censorship. Elmgren's journal notes from the 1850s and 1860s constitute one of the most fascinating cultural-historical sources for the period; they also reveal what a bitter pill it was for him to swallow that he had never been appointed head librarian.

Elmgren published in a wide variety of fields. After some translations into Swedish of various German theological treatises and writing surveys of German theological issues in the 1840s, his interests began to shift increasingly to historical and literary-historical topics. His first writing assignment dealing with literature came to him in his capacity as secretary of the Finnish Literature Society, in which he served from 1846 to 1866: he was asked to write a report on the first seventeen years of the society's operation. The society's minutes were written in Swedish at that time, since early attempts to write them in Finnish had proved an overwhelming task. In his 45-page report, *Berättelse om Finska Litteratur-Sällskapet i Helsingfors under de första 17 åren* ("Report on the Finnish Literature Society in Helsinki During the First 17 Years," 1848), Elmgren described the central role the society had played in the development of Finnish-language culture. In the massive biographical project of the 1850s, *Finlands minnesvärde män* ("Finland's Notable Men," 1853-1857), Elmgren was one of the main editors. The idea was to publish a series of volumes providing biographical information on the noteworthy men of Finnish history, especially cultural history. The series came to an abrupt halt after the publication of only two volumes; Elmgren's role in those volumes was to cover Finnish clergymen in the modern era. He ended up writing some 100 pages of the text.

Elmgren's most important work from a literary-historical point of view appeared in the early 1860s. The two-volume dissertation he wrote for a librarianship application, *Öfversigt af Finland's litteratur* ("Overview of Finnish Literature," 1861, 1865), is fundamentally an index of literature published in Finland; but Elmgren's detailed description of the books' history is so extensive and thorough that, as Kai Laitinen writes in his 1980 article on literary history, it can be considered the first comprehensive history of Finnish literature (20-21). His official defense of the second volume in late October of 1865 was a historic event in another sense as well: it was the first time women were allowed to attend a dissertation defense.[66]

Though Elmgren's scholarly interests were divided fairly evenly between Finnish cultural history and literary history, he remained one of the most prolific practical literary critics of the Cygnaeus era. His articles covered such subjects as the Estonian story of Kalevipoeg, the playwright Petrus Pachius, and late-seventeenth-century Finnish poetry. He was also entrusted with an important national task as the first editor of Porthan's works (*Opera selecta I-IV*, 1859-1873).[67]

Gustaf Edvard Ingelius (1818-1864) published his dissertation in history, *De statu Galliae Transalpinae politico et sociali usque ad tempora C. Jul.*

Caesaris (1850), and then, in an 1854 librarianship dissertation, *Froissart såsom krönikeskrifvare* ("Froissart as Chronicler").[68] According to Elmgren, who was Ingelius's schoolmate and librarian colleague (and rival for the same jobs, which may affect Elmgren's reliability), Ingelius was the special favorite of head librarian Törnegren, and consciously strove to imitate Törnegren in everything without understanding the subtler points of library science (94, 96, 108).

After Törnegren's death, Ingelius wrote a 180-page librarianship dissertation on medieval mystery plays with the imposing title: *Om Medeltidens Skådespel och deras fortgång till sednare tider, med särskildt fästadt afseende vid Finlands äldste dramatiske författare Jakob Pehrsson Chronander* ("On Medieval Drama and its Transformations in Later Eras, With Particular Attention to the First Finnish Playwright, Jakob Pehrsson Chronander" 1861). In the text Ingelius says he got the topic and a good deal of ready-to-use research material from Törnegren in 1859; Törnegren was planning to write a biography of Chronander for the unfinished collection of Finnish biographies, *Finlands minnesvärde män.* Ingelius discusses the medieval mystery plays and moralities in terms of their origins in classical antiquity, the generally religious origin of drama, the satirical and comic elements of medieval drama, the medieval history of drama, and the continuity of medieval performance traditions in various parts of Europe, such as Oberammergau. After a brief summary of the Danish and Swedish history of mystery plays and moralities, Ingelius devotes the remaining third of his dissertation to an analysis of Chronander's play *Surge.* Chronander was born in Sweden but educated in Turku, and so is often called the "first Finnish writer." Ingelius argues that *Surge* is an important cultural-historical document for study of the 1640s and artistically much more interesting than previous scholarship had recognized. Ingelius's historical overview of the mystery plays and moralities is largely compiled from other sources – notably Mone's *Schauspiele des Mittelalters*, Magnin's theater research, and Alt's *Theater und Kirche* – but nevertheless learned and well-written. The analysis of Chronander's play is commendable comparative literary criticism.

In 1854 Ingelius was appointed assistant head librarian and in 1862 head librarian, but two years after the latter appointment he took ill and went on sick leave, soon after dying in Paris.

Like Elmgren and Ingelius, Karl Collan (1828-1871) worked most of his adult life in the university library; and he too, like them, directed his scholarly interests most strongly in the direction of literary criticism, though he is best remembered today as a composer, particularly of choral pieces. Like many other first-generation Finnish literary critics, he was born the son of a clergyman, in Iisalmi. Upon completion of his Master's degree in 1850 he taught German and Swedish for a time in the Helsinki schools; starting in 1858 he worked as a librarian in the new student library; and in 1859 he was appointed lecturer in German at the university, where he lectured frequently on recent German literature. He also translated numerous German poets into Swedish, and wrote an article on the *Niebelungenlied* (1858).

Collan did most of his criticism in the gray area between folklore and literature. In his 75-page dissertation *Öfversigt af Serviens historiska folksånger* ("Overview of Serbian Historical Folk Poetry," 1860), he begins with a

Sven Gabriel Elmgren (M)

historical survey of Serbian folk poetry – awareness of it, collecting it, scholarship on it, its publication – and then devotes three chapters to the folk poems based on Stefan Drushan, Zar Lasar, and Prince Marko. He draws on Karadzic Vuk's recent edition of the poems (1841-1846), and supplements that collection, though not without some reservations regarding their accuracy, with Herder's German translations. Collan's answer to the question about the poems' historical foundation is typically romantic: "out of that material which reality offers, the folk poem from the outset selects only that part that thrills the poetic mind, that appeals to the national sensibility – in its heroes the folk personifies itself and its own honor" (9). His analysis of the poems themselves places them primarily against the backdrop of medieval Serbian history, but he insistently stresses their nationalistic aspect: behind the poems he sees the Serbians' struggle against their "hated oppressors" the Turks (63), and he reads the surviving poems as a spiritual guide for the "brave" Serbian people in their modern struggle for independence "on the road of European culture."

A similar nationalistic spirit imbues Collan's librarianship dissertation as well, *Königinhofer- och Grünbergerhandskriften* (1865), in which he publishes Swedish translations of, and comments on the scholarly debate over the authenticity of, two Bohemian manuscripts that had come to light in the early nineteenth century and were presumed of medieval origin. Both manuscripts have since been proven to be forgeries (see Mikkola 435), but Collan argues that they are based on authentic folk poems and have merely been collected and artistically "stylized" by a later poet.

In his study of Serbian folk poetry Collan also poses the romantic question of the possible existence of a lost national epic (48). In theory Collan accepts the romantic axiom that some peoples have recorded in their poetic traditions a coherent and comprehensive picture of their ancient Heroic Age; and while there is no certainty of this, the epic core of Serbian folk poetry suggests that a lost epic whole might once have existed, though not, perhaps, of the significance of the *Iliad* or the *Kalevala* (49-50).

In his second scholarly monograph, *Om bibliografiska systemer och bibliotheksmethoder* ("On Bibliographical Systems and Library Methods," 1861), Collan focused on bibliographical systems and library methods by way of qualifying himself for the head librarianship, to which he was appointed after Ingelius in 1866.

Like Elmgren, Collan was a prolific scholar who published articles both in Finland and abroad. In 1865 Collan collected some of his articles on art in *Studier och skizzer* ("Studies and Sketches"), and published it in Finland. Other important sectors of his work include serving as editor-in-chief of *Litterär Tidskrift* in 1864, developing a reader for Swedish-speaking Finns, and translating the 1835 edition of the *Kalevala* into Swedish (1864, 1868). Collan's multifaceted career was cut short by the cholera epidemic that ravaged Helsinki in 1871.

Early Criticism in Finnish

Until 1852 all scholars were required to write their dissertations in Latin; in that year Swedish was added to the short list of approved languages for scholarly writing. Once this drastic expansion had been made, it was not long – six years, till 1858 – before Finnish was included as well; and within a very brief period after that, three literary-critical dissertations were written in Finnish, by the journalist Fredrik Polén (1858), the literary and folklore scholar Julius Krohn (1862), and the Finnish language scholar August Ahlqvist (1863). Polén's dissertation was published *before* the new bylaw allowing the use of Finnish, in fact, by a special waiver; and Ahlqvist had written his doctoral dissertation nine years earlier in Swedish, in 1854, on the history of the earliest Finnish language research. The 1863 dissertation in Finnish was for the professorship in the Finnish language.

This spate of dissertations in Finnish did not however signal the beginning of a new era of Finnish-language scholarship: it would be more than thirty years after Ahlqvist before the next dissertation would be published in Finnish. The use of Finnish in these three dissertations in the five years following the bylaw change was largely polemical – it was a way of arguing for the scholarly use of Finnish – and the scholars in question had to face enormous obstacles in carrying out their polemics, both the prejudices of the scholarly community and the inadequate semantic resources of the Finnish language.

Fredrik Polén (1823-1884) was the son of an associate pastor from Eastern Finland.[69] He attended the Porvoo high school in the 1840s, while Runeberg was rector there, studied in Helsinki, and joined the editorial board of

Suometar, which was founded in 1847 to promote the Finnish language. He took his Master's degree in 1850, with a major in classics and a joint minor in history and geography. He was one of the most important journalists writing in Finnish in the 1850s, but due to some failed publishing ventures he became a teacher, first in Viipuri, then in Kuopio, in both cities continuing his journalistic activities.

Polén's *Johdanto Suomen kirjallishistoriaan* ("Introduction to the History of Finnish Literature," 1858) was the first dissertation written in Finnish. As recently as the early 1850s the Finnish Society of Sciences had pronounced unfavorably upon the use of Finnish as a scholarly language, saying that Finnish had not yet developed far enough even to become "the vernacular of the Finnish educated class, let alone a language of instruction; it is therefore not yet suitable as a communicative vehicle in the abstract sciences, in which sheer terminological problems would be insurmountable."[70] When in March 1858 the Division for History and Linguistics approved the publication of Polén's dissertation, there is a hint in Dean Cygnaeus's statement that he is not entirely convinced of the dissertation's linguistic flawlessness. His criticism is, however, touched with his usual playfulness: "Academic decorum would be served, I think, if the author were to excise one or two of the juicier bits, to digest which one must needs have an ostrich's stomach" (cited by Elmgren in his notes, 320).

Polén's dissertation was Fennophile right down to the title page: he had Finnicized his first name, rendering Fredrik as "Rietrik." Elias Lönnrot, who chaired the dissertation defense, argued strongly in favor of the scholarly use of Finnish and launched a pointed attack on those who would carp at or dismiss Polén's project:

> But it is already beginning to seem as if their grimacing and griping, their sneers and their slanders, their public and private machinations and opposition, in a word all their conniving no longer has much of a future. In the Finnish fatherland the Finnish language will triumph whether they like it or not. If they cannot accept it with good grace, let them accept it with bad, or else let all such xenophobes remove to whatever they think their true homeland to be.[71]

In his 79-page dissertation Polén seeks to construct a framework for the history of Finnish literature by exploring the Finnish language's readiness for scholarly use, the nature of scholarly research in general, the concept and significance of literature, and early work on the history of Finnish literature; he also attempts to periodize Finnish literature, to break its history down into distinct historical periods. Polén insists strenuously on the writing of Finnish literary history in Finnish, no matter how difficult the beginning may be.

> At the present Finnish literary history might be compared to the agricultural process of burn-clearing. I plow the rough soil, jarring over rocks and stumps;[72] let later and greater critics harrow and smooth. To everything its time and craftsman. Let one stumble, and another will right him – or rather, science itself, in its rational growth, will right him. The sciences obey the eternal laws of reason. (2)

Polén argues for the division of Finnish literary history into two main eras, the ancient and the Christian. In the former Finnish literature was linguistically purely Finnish, and this era grounds and shapes all subsequent literary periods: "Through it the Finnish people have reminded every living generation, in particular its educated citizens, as if to say: 'Here I am – do not forget me, nor forget yourselves.'" The ancient era for Polén was monolithic; the Christian era divides up further into periods bounded by Mikael Agricola's Finnish primer in 1543, the founding of the University of Turku in 1640, the formation of the Aurora Society in 1770, and the founding of the Finnish Literature Society in 1831. That last date ushered in a period that Polén characterizes as "the dawning of Finnish national literature, whose brilliance has year after year sharpened and taken on clarity until the present" – though much work remained to do before "culture in Finland" was ready to follow "its very own unique path."

Polén's periodization reflects a vision of cultural history according to which the history of literature must follow the stages and directions of a nation's cultural development in general. In fact, Polén polemicizes against Rancken's 1851 dissertation; Rancken had drawn up a period framework for the history of the Finnish book with the founding of the university, the Uusikaupunki truce (1721), and the Hamina truce (1809) as period boundaries. The two last Polén considers mere external events that had nothing to do with the development of the spirit. Polén's dissertation is steeped in the romantic idea of national individuality, which he got from Gervinus, as also in Hegel's concept of history. Polén refers several times to the Swedish literary historian Peter Wieselgren, whose history of Swedish literature was well-known in Finland; in fact, Polén takes his epigraph from it: "If we scorn our own language, our own literature – then Europe will scorn us." Polén's most important influence, however, seems to have been Snellman. Paraphrasing Snellman, Polén calls literary history the "inward side of national history" and "the inward conscience that awakens the people" (78-79).

Polén's dissertation has its obvious scholarly failings. His sources are haphazard, his discursive style is essayistic, he digresses constantly, and the problem of critical terminology in Finnish looms large. Certain of Polén's contemporaries, including the Fennophile Elmgren (320) and the historian Yrjö Koskinen, who defended his Finnish-language dissertation later that same year, called the dissertation a scholarly fiasco. Still, his periodization of Finnish literature was not without influence; several later scholars picked it up and refined it. Polén himself seems to have taken the adverse criticism his dissertation aroused hard, for he stopped doing scholarship, leaving the promised sequel to the dissertation, on the actual history of Finnish literature, to other scholars. The first to continue in the direction Polén pointed was Julius Krohn.

Krohn (1835-1888) was active on many fronts in the development of Finnish-language culture.[73] This was perhaps somewhat unexpected, since his family – and thus his first language – were German. Krohn's father came of St. Petersburg German stock (later of Viipuri); Krohn did not learn Finnish until he was at university. Still, in his language politics he was a fervent Fennophile. His programmatic Fennophilia is evident in his motivation for his earliest scholarly work, the Finnish-language dissertation (written for a docentship) mentioned earlier, *Suomenkieleinen runollisuus Ruotsinvallan aikana ynnä kuvaelmia suomalaisuuden historiasta* ("Finnish-Language Poetry During Swedish Rule, and Some Depictions of the History of Fennicism," 1862). In his own introductory words, he intends to demonstrate the centuries-long roots of the Finnish language and the Finnish nationalist spirit. When his dissertation was completed, he wrote to his parents:

> The dominant idea that unifies the text is one I have already explained to you at various times, namely to present and prove that entirely neglected fact that under Swedish rule the Finnish people were not without nationalist spirit, and that the Finnish language even after 1700 was the general language of social intercourse in all segments of society. . . . I feel gladdened by my sure hope that my discovery of the above-mentioned fact might have the effect of strengthening and uplifting the spirit of nationalism in our country. For it is evident that Fennicism is not yesterday's invention, a passing illusion grasped at by those still wet behind the ears, but rather a long tradition that was merely relegated for a time to the periphery. (Cited in Trast, 44)

Krohn's dissertation can be considered the foundation for his later literary-historical studies. His material for the 183-page dissertation was the body of Finnish-language poems and hymns written under Swedish rule – an imposing number of texts largely collected by Matti Pohto (1817-1857) and stored in the University of Helsinki library. Pipping's bibliography provided Krohn with another significant source; without it, Krohn says, "it would hardly have been possible to undertake a large-scale project like this, the writing of a comprehensive history of Finnish literature." Krohn's analytical method is largely one of describing the literary texts, rather than seeking historical explanations.

The dissertation qualified Krohn for a docentship in Finnish language and literature. Later he dedicated himself wholly to an academic career: he was appointed lecturer in Finnish in 1875 and in 1885, three years before his death by drowning in a sailing accident in the Bay of Viipuri, he became professor extraordinaire of Finnish language and literature. Previous to this appointment he had deputized in the professorship for years, while its incumbent, August Ahlqvist, was on leave of absence.

Krohn's dissertation was followed by a ten-year hiatus in publication, however, and his most significant scholarly work did not appear until the 1880s. The cultural imperative of the mid-nineteenth century was to write a history of literature in Finnish, but after the preliminary work done by Polén and Krohn, the writing of a comprehensive history of Finnish-language literature had to wait another quarter of a century. There were two main obstacles. One was that it was difficult to write the history of something that hardly existed (in the modern sense of the word "literature") in sufficient quantities yet; the other that it was difficult to undertake the project without adequate bibliographies of Finnish-language literature. The latter obstacle was removed in the 1870s by Vasenius, with the publication of his comprehensive bibliography of Finnish-language literature (supplementing Pipping's index); Krohn himself referred to the former problem in 1885, in the first volume of his projected three-volume history of Finnish literature:

Julius Krohn with his daughter Aune (SKS)

> In beginning this story of the development of Finnish-language literature, I am well aware that it will meet with a skeptical curiosity on many hands. Whereof will I speak? Where is that Finnish literature about which a book might be written? These questions will no doubt rise in many minds as readers pick up this book. (1)

Krohn's task is facilitated, however, by his broad national-idealist theory of literature, according to which Finnish-language literature includes all literary products "created" by the Finnish-speaking people, including the *Kalevala* and the collection of folk poetry *Kanteletar*, each of which was assigned its own volume in Krohn's planned work. By the time of his death in a sailing accident Krohn had only brought out the first volume, on the *Kalevala*; the second and third volumes were edited and published by his son, Kaarle Krohn (1863-1933), himself a folklorist. In his foreword to the third volume, *Suomalaisen kirjallisuuden vaiheet* ("The Stages of Finnish Literature," 1897), Kaarle Krohn notes that his father had left the unpublished second and third volumes of the series in a state that he, Julius, had considered almost completed, "lacking only a few finishing touches"; but editing the third volume, Kaarle Krohn goes on to say, was "like the compilation of redactions" (vi).

The first part of *The Stages of Finnish Literature* is based on the lectures Julius Krohn delivered in the mid-1870s at the Helsinki Academy for Women.

The lectures extended in time up to the early nineteenth century, to the work of Karl Axel Gottlund, an early collector of folklore and himself a writer – or roughly to the middle of the period covered by the volume. Before these lectures Krohn had lectured on Finnish literary history at the university, as a docent in Finnish language and literature in the 1865-1866, 1871-1872, and 1872-1873 academic years; but Kaarle Krohn had to go to the *Biografinen Nimikirja* ("Biographical Register," 1879-1883) for the written material that made up the second part of his third volume. Some material was also available from his father's lectures in 1885-1886, when Julius had covered Finnish literature up to the present. In any event, the first part of this third volume is considerably more coherent than the second, which the editor was forced to compile from exceedingly diverse materials.

The Stages of Finnish Literature is a 480-page volume that covers extensive literary-historical ground from the Reformation to the 1880s. Krohn's periodization follows current cultural-historical practice in breaking periods or "stages" off at critical events in the history of the use of Finnish. The earliest or "Reformation period" begins in the 1540s and ends with the foundation of the University of Turku in 1642. The second, or "Fennophile period" – one of Krohn's terminological contributions to Finnish literary historiography – ends in 1809 with Finland's removal from Swedish rule. The third, which Krohn calls the "nationalist or Fennomania period," is split in half: first comes the "battle-of-the-dialects era," then, following the critical period of Snellman's nationalistic journalism in the 1840s, the "new Finland era."

Krohn's history of Finnish literature is somewhat programmatic in its polemics for the organic unity of literature written in Finnish and Finnish national existence. The Finnish Reformer Mikael Agricola, for example, is characterized by saying that his Finnish Bible launched Finnish literature and thus saved Finland "from disappearing off the face of the earth" (30). Krohn then traces the subsequent battles of Finnish against the dominant Swedish language and culture, concluding in the 1880s with the situation in which "the Finnish nationalist cause" had so shaped the thinking of the intelligentsia that "its outward form, even its language, is being Finnicized" (473). Krohn devotes considerable space to a whole range of minor Finnish writers, including religious, linguistic, journalistic, and scholarly writers and translators. He passes over the work of the major nineteenth-century Swedish-language authors – Runeberg, Topelius, and Wecksell – with hardly a word, and gives only a very brief account of Snellman's work.

The theory that national development is language-bound and folk-autonomous was already outdated in the 1880s. This was clear in the late-nineteenth-century reception of Krohn's work. Reviewers argued that Porthan's scholarly work and the period it dominated could only be fit into a "Fennophile" period as into a Procrustean bed (Roine, A. V. Forsman, Grotenfelt); they criticized Krohn for neglecting Swedish-language literature, particularly the nationalistically influential Saturday Society and Snellman (Forsman); and they attacked the notion that Finnish literature had developed in an idealized state of national autonomy, arguing that Finnish literary history from the beginning reflected modifications of large intellectual and cultural trends sweeping Finland from abroad (Saxén).

In any event, Krohn's three-volume work was the first extensive history of

Finnish literature. And despite the criticisms launched at its nationalistic and linguistic chauvinism, in practice it bridged the nineteenth-century nationalist literary-historical tradition into the twentieth century. Indeed one of Krohn's reviewers, A. V. Forsman, agreed with Krohn to the extent that the history of Finnish literature is "on a large scale identical to the history of the Finnish people's intellectual cultivation" – signifying by "Finnish people" the Finnish-speaking population. Rafael Koskimies, one of our own century's greatest Finnish literary historians, continued this tradition during the Second World War and after in his own three-volume history of Finnish literature, *Elävä kansalliskirjallisuus* ("Living National Literature," 1944-1949), which was subtitled "The Development of the Finnish Spirit 1860-1940."

Krohn's literary history was preceded by pathbreaking research in the field. In 1876 he had been appointed chairman of the Hymnal Revision Committee, which had at long last, after numerous miscarriages, been formed to modernize the 1701 edition of the Lutheran Hymnal. Upon completion of the committee's work, Krohn published the 100-page *Suomen Wirsikirjan Historia* ("A History of the Finnish Hymnal," 1880) in which he traced the origins of the existing hymns and the birth of the suggested revision. In his *History* Krohn places particular stress on the linguistic form of the hymns and argues strenuously for modernization, against all those who claimed that to modernize an old beloved hymn was to corrupt it.

In 1850 the university had created a professorship in the Finnish language; when the first appointee M. A. Castrén died in 1852, the title was changed to "professor of Finnish language and literature." During the tenure of Castrén's two successors, as Mikko Korhonen has argued, Finnish literature and folklore were vaguely seen as falling into the professor's purview. Elias Lönnrot (1802-1884), for example, whose 1827 dissertation had been *De Väinämöine priscorum Fennorum numine*, lectured in the 1850s on the *Kalevala* – though his lectures were largely Swedish annotations to the difficult Finnish text. Lönnrot was no literary scholar, but his work proved significant for the study of literature in at least two ways.[74] In the first place, in a long-running debate on the formal prerequisites for Finnish lyric poetry, he had argued strongly for the construction of the Finnish lyric on the foundation of the folk poem. In his foreword to *Kanteletar* in 1840, he suggested that "Finland's new song be raised on the cornerstone of the old one," i.e., that the traditional folk poetry serve as a formal model for new Finnish poets. As evidence for the incompatibility of foreign poetic forms with Finnish lyrics, Lönnrot offers the fact that the Finnish people have not taken the poetry of learned poets to heart: these writers have "fallen away from the mind, the customs, and the familiar songs of the people" (l-li; cf. Haapanen). In his 1845 publication of Runeberg's poems in Finnish Lönnrot appended an explanation of their metrical form – the first sketchy study of Finnish metrics to be written. The novel aspect of Lönnrot's metrics was that they were based on quantity.[75]

During his tenure as professor of Finnish language and literature Lönnrot gave the first lecture course ever to be held in Finnish at the university, and, ever the prolific wordsmith, he coined a remarkable number of new literary-critical terms. These include *kirjallisuus* "literature," which first appeared in 1831 in the name of the Finnish Literature Society; also *romaani* "novel," *runoelma* "poem, poetic work," *runomitta* "poetic meter," *kirjoitelma* "essay, piece," *kuvaelma* "scene, tableau (vivant)," *laulelma* "tune, ditty," *lukemisto* "anthology, reader," *muinaisruno* "ancient poem," *itkuvirsi* "lament," *kertomaruno* "narrative poem," *toisinto* "redaction, version," and *julkaista* "publish."

Lönnrot's successor in the professorship of Finnish language and literature was far more significant a literary scholar. He was August Ahlqvist (1826-1889), best known as a poet, linguist, reviewer, and tireless vilifier of Aleksis Kivi; but alongside Julius Krohn he was the first important literary critic who wrote in Finnish.[76]

Ahlqvist's first literary study was *Viron nykyisemmästä kirjallisuudesta* ("On Modern Estonian Literature," 1855), born out of a trip Ahlqvist made to study the Russian Finno-Ugric tribes – during the trip Ahlqvist spent some time in Tarto. For decades, until the end of the century, this book and D. H. Jürgenson's *Kurze Geschichte der Ehstnischen Literatur* (1844) were the only two histories of Estonian literature extant.

Ahlqvist's central literary-critical work was his professorial dissertation, *Suomalainen runous-oppi kielelliseltä kannalta* ("A Linguistic Approach to Finnish Poetics," 1863). In his introduction Ahlqvist defines poetry, referring to Rudolph von Gottschall, in a somewhat more formal way than the romantics: a weighty content is not enough, he says, to make an artistic impression; only form "charms and entrances" the reader. These linguistic requirements vary slightly from language to language, but it is indispensable that "the words in poetic language not follow one another haphazardly but obey the laws of decorum, and sound sweet to the ear." Ahlqvist then examines the meter, rhyme, and repetition of Finnish folk poetry, and notes that Finnish poetry is currently shifting into new European meters: this, he says, "is not to be regarded as a fault but as progress." He also offers his own terminological suggestions: he is not satisfied with *kirjallisuus* "literature," for example, which was by then established in the language. He argues that it is as illogically formed as *runollisuus* "poetry," which should, he says, be *runous*: a form that did in fact displace the older *runollisuus* and become the Finnish word for poetry. He was not as successful at displacing *kirjallisuus*; but other coinages he suggested did stick, including *juoni* "plot," *näytelmä* "play" (in the sense of drama), *runojalka* "poetic foot," and *runousoppi* "poetics."

Ahlqvist's main theoretical authorities were Hegel, Vischer, and above all Gottschall, whose *Poetik* (1858) became his aesthetic Bible. In his negative review of Calamnius's Finnish translation of Aristotle's *Poetics*, he insisted that it would have been more appropriate to translate Gottschall's *Poetik*.

Though Ahlqvist was first and foremost a linguist, he too regarded the creation of a Finnish literature one of the most important cultural tasks facing the Finnish people. He argued, however, that Finnish literature could not be produced until the Finnish literary language had been developed – until, as he put it, the linguistic "field" had been cultivated by "less-gifted ditch-diggers and hoers" like himself. In this light it is understandable that, as Kohtamäki points out in his biography, Ahlqvist should appoint himself the "merciless gardener" of the Finnish language. To his mind literature was a "new field," a "meadow," or a "garden," the critic a "gardener" and an "overseer," reviews "weed-killers." Ahlqvist sought to perform this social role in his *Kieletär* magazine.

Ahlqvist's aesthetic Puritanism seems to have stemmed from this self-appointed role as linguistic policeman, for the most interesting piece he wrote on aesthetics indicates that he had unmistakably romantic inclinations. In the opening statement he read at the defense of his professorial dissertation, "Runouden raaka-aine" ("The Raw Materials of Poetry," not published until 1897), he listed the subjects that in his estimation (which he borrowed almost verbatim from Gottschall) were appropriate for poetic use. Nature, for example, was suitable when it served as a backdrop for human mental states:

> Thus for example a craggy mountain slope, up which boulders climb one above the other without end toward the sky, may well serve as foundation for a poetic tableau in which the poet seeks to depict the struggles of the human heart and movement of the human mind. By the same token the poet may well paint the sparkling surface of a lake in the summer morning sun, reflecting the leafy-green islands and shores and almost resounding with the tinkling of cowbells. A vista of this sort provides the ideal foundation for a description of that calm, placid, contented state of mind that is at peace with itself and the world around it. In general, poetry will do well to employ images and symbols from nature in portraying the movements and stases of the human mind; particularly such forces of nature as thunder, storm, sunrise, dusk, etc., are appropriate for this use.

Elsewhere in this statement, in discussing the suitability of various historical periods for poetic use, Ahlqvist had his first opportunity to attack Aleksis Kivi, who had just won the Finnish Literature Society play competition with his *Kalevala*-based play *Kullervo*. Paraphrasing Gottschall, he claims that the oldest or "ancient" historical period is not suitable for dramatic use, and then uses the recent dramatizations of Finnish folk poetry by Brakel and Stenvall (Kivi's real name) as examples of this unsuitability. It would not be possible to stage ancient history successfully, even "with a greater poetic talent and a stronger poetic skill than these men have shown."

The best period for literary use is the present, with "all its activities, hopes, attractions, and errors"; but even the present must be portrayed from a perspective that prevented the fall into "insignificant mundanity." Here Ahlqvist's notions hark back through Gottschall to the German theory of "romantic realism" offered by Otto Ludwig. Both German writers would later offer Ahlqvist theoretical support for his virulent attacks on Aleksis Kivi. According to Gottschall "a reductive realism produces a work of art dominated by lifeless nature," and "the rise of realism as a dominant stylistic principle could only lead poetic art astray" (quoted in Marckwardt 327-328). In was on precisely this sort of ground that Ahlqvist panned Kivi's *Seven Brothers*.

In his professorial capacity Ahlqvist inclined most strongly toward linguistics; indeed his repeated lectures on the *Kalevala* were linguistic in method. On the other hand, he had lectured on the Finnish poetic meter as a docent in the autumn term of 1860, and it is reasonable to assume, as Simelius suggests (81-82), that his choice of Finnish poetics as the topic of his professorial dissertation was indicative of his conception of the field. In two of Ahlqvist's polemical pieces during his tenure as professor he also touches on literary-critical issues. "Uusin Suomalaisen Virsikirjan ehdotus kielelliseltä ja

August Ahlqvist (M)

runolliselta kannalta tarkastettuna" ("The Proposed New Finnish Hymnal Examined from a Linguistic and Poetic Standpoint," 1884) is a critique of the hymnal committee's work in which, partly ironically, but in any case with a good knowledge of the Finnish language, Ahlqvist points out the linguistic inconsistencies and inaccuracies in the proposed hymnal revisions. He appeals to the people's feel for their own language, and in many cases is more willing than the committee to stick with an old familiar phrasing. According to Ahlqvist enjambment is inappropriate in Finnish hymns, because it is not used in Finnish folk poetry. It is somewhat surprising that Ahlqvist, who came from Eastern Finland, should criticize the committee's attempts to distance themselves from the Western Finnish tenor of the old religious register and introduce some Eastern Finnish expressions into the hymnal; one explanation may be Ahlqvist's and Krohn's scholarly disagreement over the regional provenance of the *Kalevala.* Ahlqvist claims that Krohn's committee has provided an inadequate foundation for a revised hymnal; a better one, he argues, would have been Lönnrot's alternative proposal, *Suomalainen Virsikirja väliaikaiseksi tarpeeksi* ("An Interim Finnish Hymnal," 1872) and its revised edition (1883), since for linguistic reasons both better satisfy Finns' "spiritual needs."

Ahlqvist's *Kalevalan karjalaisuus* ("The Karelian Origins of the Kalevala,"

1887) was another polemical response to Krohn; we shall return to examine that text below.

The late 1850s and early 1860s marked a significant expansion period in the development of Finnish-language literary criticism, with progress in the field running concurrently with the early work of the first great Finnish-language writer, Aleksis Kivi. After the 1860s literature in both Finnish and Swedish lapsed into a dormant phase that lasted until the 1880s. In literary criticism, Cygnaeus's successor C. G. Estlander mediated between the first and second generations of Swedish-speaking critics, and on the Finnish-speaking side both Krohn and Ahlqvist continued publishing after the 1860s; but a new Finnish-language literary criticism was a long time in the making. Part of the reason for this, again, was the underdeveloped state of scholarly Finnish; but more important than that even was a shift in the general cultural climate, one that was not conducive to the expansion of Finnish-language scholarship.

Other Critics

There were in this period independent critics as well, who worked outside the university, primarily in schools or the press. The most significant of these in the second half of the nineteenth century were Johan Oscar Immanuel Rancken and Gabriel Lagus.

Rancken (1824-1895) spent his adult life as a high school teacher in Vaasa, first, beginning in 1846, as a lecturer in German, then in history and geography, and finally, beginning in 1865, as rector of the school.[77] Rancken was the son of the verbally gifted pastor Engelbrekt Rancken, who in the 1810s had translated Goethe's *Torquato Tasso* into Swedish and later edited a church periodical that followed German cultural life closely.

Rancken's scholarly interests covered a broad spectrum of history, literature, and ethnology. He published a dissertation on sixteenth-century Finnish history in 1850, and another on historical and geographical literature published before 1722, prior to the so-called "Great Hate" (*De litteris historicis et geographicis Fennorum usque ad academiam Aboensen anno 1722 restitutam*) in 1851. The latter is little more than a list of books, including literature published by Finnish writers either in Finland or abroad. Rancken's approach is purely bibliographical; he attempts no historical explanations of any kind. According to Törnegren, who served as official faculty opponent, Rancken dealt more with the books' outward features than with their content; in addition, his classifications were not entirely consistent.

Rancken's interest in ethnology is evident in his 1859 book *Folksången hos de romaniska och germaniska nationerna I* ("Folk Songs in the Romance and Germanic Nations") and his 49-page librarianship dissertation *Cid i de spanska romanserna, hos Corneille och Herder* ("El Cid in the Spanish Romances, as Treated by Corneille and Herder," 1861). In these two Rancken is methodologically shaky; he is clearly more of an enthusiastic amateur than he is a solid and profound scholar. The former work was intended as the first volume of a series on the history of European folk poetry, but no further volumes were forthcoming. Rancken combines information gleaned from Herder and Bouterwek with his own observations on French, Provencal, Spanish,

Portuguese, and Italian folk poetry, and explores the work of late-medieval art poets as well, wherever they drew on folk poetry. He uses much the same analytical method in his study of El Cid: he examines the Spanish El Cid romances, Corneille's *Cid*, and Herder's German translations of El Cid; runs rather haphazardly through the secondary literature on El Cid; then concludes with his own Swedish translation of El Cid, which had not been published due to Oscar Fredrik's untimely publication of his own Swedish translation in 1859. In his opponent's statement Cygnaeus notes that Rancken's dissertation offers nothing new, and on his recommendation the division rejected the dissertation in the spring of 1862.

In his work on the history of Finnish literature Rancken preferred to study forgotten and relatively minor writers like the poets Mikael Choraeus and Jakob Gabriel Leistenius. Rancken's little monograph *Mikael Choraeus* (1875) did unearth new archival material on the poet's life and a few previously unknown poems. In a slim volume on *Axel Germund Lilljegren* (1887) Rancken introduces the reader to a late-eighteenth-century occasional poet, providing biographical data about the former soldier and reprints some of his congratulatory and other occasional poems. The studies Rancken wrote of Finnish literary history can best be read as a late-nineteenth-century continuation of the tradition begun in the history of learning: one introduced the reader to old texts, in this case texts from the period of Swedish rule.

In his last work, *Om J. W. v. Goethes skådespel Torquato Tasso i öfversättningar till svenska* ("On Goethe's Play *Torquato Tasso* in Swedish Translations," 1888), Rancken discusses his father Engelbrekt Rancken's early Goethe translation and the public debate that surrounded it.

Beginning in the early 1850s Rancken tried over and over to get a job in the university library, applying for an assistant librarianship in 1854 and the head librarianship, left vacant by Törnegren's death, in 1860. He never did get one. If we are to believe Elmgren's journal notes, Rancken's scholarly qualifications were not highly regarded.

Gabriel Lagus (1837-1896), too, the scion of a famous scholarly family, made his living outside the university.[78] He was a considerably more important scholar than Rancken, and of the first-generation literary critics he was most strongly grounded in literature's historical context. The year after completing his Master's degree, in 1858, Lagus published a dissertation on Greek geography and modern history. This qualified him for a high school teaching job in history and geography, and in 1859 he got one in Porvoo; in 1867 he moved to Viipuri, where he taught Greek and Hebrew. Lagus did not think of teaching as his professional calling, however, and all his life worked outside of his teaching job as a journalist, playwright, historian, and literary critic. His journalistic pieces, in particular his essays and reviews, were highly influential and broke considerable new ground for later cultural criticism.

In his first literary-critical study, his 69-page dissertation *Gustaf den tredje och hans tidewarf, uppfattade af poesin* (1860), Lagus discussed Gustaf III and his period as a literary motif. The texts he deals with are fairly recent: they include Topelius's play *Efter 50 år* ("After 50 Years") and three Swedish works, Philipp Meyer's plays *Året 1772* ("The Year 1772") and *Året 1789* ("The Year 1789") and K. A. af Kullberg's novel *Gustaf III och hans tid* ("Gustaf III and His Time"). This last Lagus rather uncritically calls one of the greatest accomplishments in the history of the Swedish novel and far and away the greatest Swedish historical novel.

In his dissertation Lagus makes no reference whatever to secondary literature – in this sense it is more like a long essay than a piece of scholarship. It is, however, solidly grounded in literary theory, specifically the complex theoretical relations between literature and reality, genre theory, and period types in characterization. In his introductory chapter Lagus explores the philosophical problems of the individual vs. the collective and freedom vs. necessity as a background for his discussion of the "Gustavian period." These problems he takes to be the essence of all true tragedy, and endemic to being. As long as the creation has existed, so too has the tragic. The individual is never content to be a mere part of a whole, but seeks to become the focal point – and this is the basic situation of tragedy. A prime expression of tragedy is found in the myths of the Bible (Jesus and Satan in the wilderness), of ancient Greece and Rome (Prometheus), in folklore (Kullervo), and in modern literature (Shakespeare, Runeberg). This is the battle between the creator and the creature, which is also the antinomy between freedom and necessity. Human freedom requires as its backdrop divine necessity; humans at their best are always guided by, and act as executors of, the divine idea. Otherwise freedom is arbitrary. A free being must follow the lead of morality; if he doesn't, he will in the end be compelled to admit that his so-called freedom has made him unfree. In addition to the divine totality, the individual must take into consideration other factors that restrict his freedom as well. The individual is the child of his country, belongs to a family, is tied to the developmental stage of his culture, and his behavior is regulated by gender, character, and the like. Is therefore the individual free at all? Lagus asks. In his view the answer must be both yes and no. The tragic subject is both free and unfree. For Lagus Kullervo is a typical example of the tragic hero: his freedom has murdered his freedom.

In all this Lagus is very much of Cygnaeus's camp. He is likewise close to Cygnaeus, particularly the dissertation on King Erik, on the issue of historical periods. Lagus argues that every period generates its own essence or nature, and the more conflictual that essence, the more capable it is of expressing the inner content of the "human heart." All transitional periods or crises are rich in conflicts, as was evident in the "Gustavian" eighteenth century. The essence of that period was the intensification of "liberated passion" into a guiding principle. Attempts were made to smash tradition. To what end did man invoke divine guidance, when he had his own reason at hand? Why should the state and all worldly existence be directed by a single transcendental idea, when a foundation and principle could be derived from worldly things themselves, from their own utility? What Lagus seems to be doing, in fact, is taking Cygnaeus's ideas about historical periods and the suitability of Swedish kings for literary use one step further: Gustaf III lived in dialectical relation with his period, and thus drew the rays of his period to him as to a focal point, thus making it possible to study an entire world view (*världsriktning*) in him. It is in this sense that he offers the artist excellent material.

Lagus devotes the lion's share of his textual analysis to Meyer's plays, even though he does not consider them of much artistic significance. His low estimation of Meyer's work is symptomatic of his whole approach to literature.

He argues that the plays are too slavishly bound to the historical record, which stunts their development both of artistic effect and of historical spirit or *Geist.* "In them the true content (*andemening*) of the period has not been grasped," and thus the characters portrayed are not "historical" in the deepest sense of the word: no matter how closely they correspond to their historical models, they lack that spiritual core that is the key to all history. They are not expressions of "one particular trend in world culture" (53).

In his discussion of Kullberg's novel, Lagus again stresses the importance of examining each character in dialectical relation with his or her own time. The time has produced the character, and only against the backdrop of time can the character be understood in its full reality. Gustaf III would remain a mere oddity were he not viewed as a product of the eighteenth century. Nor may one stop there. The whole truth about a period like the eighteenth century cannot be adequately depicted through the medium of "epic" or prose fiction; one needs drama as well, for only drama can consolidate its inner contradictions in a higher unity. In brief, the eighteenth century must be imagined as a tragedy, and Lagus argues that this was Kullberg's idea as well; unfortunately, Kullberg only partially instilled this idea into his novel.

In his last chapter Lagus fleshes out a portrait of Gustaf III and his time as a five-act tragedy, in which "the loose ends are tied up, but also judged and ultimately reconciled." In this Lagus weaves literary and historical reality in a complex fabric that renders literary criticism and theory a form of social criticism.

Lagus's magnum opus is the two-volume, 300-page history of Swedish-language Finnish literature, *Den Finsk-Svenska Litteraturens Utveckling I-II* ("The Development of Finnish Literature in Swedish," 1866, 1867), which begins with the founding of the University of Turku in 1640 and breaks at the end of volume one with Runeberg. It is written in essay style, in the form of ten "lectures," each of which forms a semi-autonomous whole that nonetheless ties up with the lectures before and after it.

Lagus is clearly familiar with the old literary documents; while the modern reader would prefer to have rather more information on writers, book titles, and publication dates, overall the work is a reliable guide to Swedish-language Finnish literature in the seventeenth, eighteenth, and early nineteenth centuries and its contextualization in cultural history. In the debate then raging over whether Swedish-speaking literature was to be regarded an integral part of the Finnish national literature, Lagus takes an unhesitatingly affirmative stance. This places him in the opposite camp from Snellman, and in fundamental agreement with Snellman's promising protege who died young, Robert Tengström (1823-1847), who argued in an article on "Teckningar från den fosterländska Vitterhetens område" ("Notes on the Realm of Patriotic Literature," 1845) that literature written in Swedish in Finland might also be Finnish-national in character. In the preface to his literary history (which was dedicated to Fredrik Cygnaeus, to whom Lagus acknowledged his great debt in the "science of literature" [*i litteraturens vetenskap*]), Lagus writes:

> It was not out of patriotism that I undertook the study of our early literature, but rather out of a sympathy for the forgotten deceased. Furthermore, I began my research on these writers with no precon-

Gabriel Lagus (M)

ceived aesthetic or political theory. As my work progressed, however, I began to sense that the individual works I was studying were all united by an invisible bond that skipped forward in time. It felt to me as if the mighty spirit of history had found self-expression in these forgotten and despised writings – that they too might be seen as evidence of our country's struggle toward consciousness. Was I right or wrong in so feeling? The answer lies in this study.

Lagus was the first to show that Swedish-language Finnish literature formed its own coherent and independent tradition that diverged markedly from that of Swedish literature. In Lagus's view Finnish literature written in both Swedish and Finnish reflect the essence of the Finnish people. The peripheral phenomena with which he begins his book, the occasional poetry and sermon collections of the seventeenth century, are valuable as evidence of "the development of culture in the midst of our poor and long-suffering people." When he deals with later, more significant poets, he goes to some lengths to establish their national Finnish character. Thus for example the early-eighteenth-century poet Jacob Frese, who was born in Finland and later moved to Sweden, wrote poems in the typically Finnish elegiac, melancholy mode that Lagus argues differentiates his work from that of his Swedish contemporaries. Lagus concludes his study in the 1830s, "in a period when, having become through Runeberg a powerful force in the country, Swedish song in Finland began to exert its strongest influence over the people." Though Swedish-language poetry had originally been "foreign" in Finland, it had gradually taken such deep root in Finnish conditions that it had become "a true expression of the people's feelings and ideas."

The work's critical reception was mixed. In a review in the liberal Swedish-language daily *Helsingfors Dagblad*, Anders Chydenius found Lagus's study a powerful and profound analysis of the development of Finnish culture – though he did not entirely agree with Lagus's belief in "an internal, primeval folk genius that finds expression in individual phenomena." The anonymous *Hufvudstadsbladet* reviewer, on the other hand, found fault with Lagus's separation of literature written in Swedish off from that written in Finnish. A Finnish literary history restricted to a single language, he argued, would always be one-dimensional, and thus "lacking in greater historical value." As might be expected, the attack from the Finnish-speaking camp was virulent. In the Fennomania party paper, *Kirjallinen Kuukauslehti* ("The Literary Monthly"), Julius Krohn bitterly criticized the exclusion of Finnish-language literature in favor of Lagus's "own" Swedish, "which swells up so mightily that it pretends to be head of the household, of which in reality it owns only a tiny corner."

The language question raised first by Robert Tengström and then, more exhaustively, by Gabriel Lagus, was later to become one of the key issues of Finnish literary historiography. Apart from Werner Söderhjelm's insistence in the early twentieth century that Finnish literature written in Swedish is just as Finnish as that written in Swedish – precisely Lagus's point – for almost a century Finnish literary histories were invariably segregated by language. Until the 1950s the dominant Finnish-language literary historians paid little or no attention to literature written in Swedish – the few exceptions, like Runeberg and Snellman, were considered of unusual "national" significance – and the minority Swedish-language literary historians were left to protect literature written in Swedish from national forgetfulness. Even then, the greatest literary achievements in Swedish were often seen as *Swedish* achievements, part of the literary history of Sweden.

PART THREE
From National Idealism Toward a Positivistic Science of Literature

C. G. Estlander and the 1868 Appointment

Fredrik Cygnaeus's successor Carl Gustaf Estlander (1834-1910) was born in Swedish-speaking Ostrobothnia. His father, Jacob Jonas Estlander, was the vicar of Lapväärti, and had earlier served as docent in the history of learning at the University of Turku; the family admired the Gustavian period and Franzén and had no great interest in romantic literature. Estlander attended school in Vaasa, where his high school principal was the Fennophile poet and theologian Lars Stenbäck; graduating from high school in 1850, he entered the University of Helsinki in 1851 with the vague idea of majoring in mathematics and the natural sciences. With the 1852 reform of the university curriculum, Estlander switched to the humanities, studying modern languages (Spanish, French, and English) under Cygnaeus, until in 1854 at his teacher's urging he focused in on literature and aesthetics.[79] Estlander seems to have made a special impression on Cygnaeus at an early stage, since in 1856, when Estlander had just completed his Master's degree, Cygnaeus appointed him official faculty opponent for Alfred Edman's aesthetics dissertation *Hegels lära om konstformerna* ("Hegel's Theory of Art Forms").

In his memoirs Estlander recalls having to spend so much time listening to lectures and reading books on literary history that there was not a great deal of time left for actually reading literature. His textbooks included Gervinus's *Geschichte der poetischen Nationalliteratur der Deutschen*; a history of eighteenth-century literature by Hettner, probably the *Geschichte der englischen Literatur* (1855), the first volume of a multivolume study Cygnaeus seems to have recommended he read; Demogeot's history of French literature; Karl Rosencrantz's history of medieval literature; and some book by the German scholar Friedrich Bouterwek, most likely the classic *Geschichte der Poesie und Beredsamkeit.* His central aesthetics text was Vischer's *Aesthetik*, which Berndtson had introduced to his students in his professorial lectures back in 1853.[80] Estlander says he had to familiarize himself with English literature without a textbook, which was, he recalls, all to the good.

Estlander also familiarized himself during his student days with the ancient poetry of the European peoples – the *Poema del Cid*, the *Song of Roland*,

Ossian, Percy's edition of old English ballads, the *Volsungsaga*, and the *Niebelungenlied* – though these were not in the course requirements. Estlander recalls in his memoirs (234, 288) that the *Kalevala* had sparked his interest in these old texts. He also studied the Finnish language, for ideological reasons: it was "the language of the people, the language that Runeberg had described in the *Elk Hunters*, and that Lars Stenbäck had taught us to love and respect in school. It was a despised, undeveloped, and neglected language, and precisely because it was base I chose it, out of a sympathy for the underdog and a desire to go against the grain – both attitudes that have followed me all through my life" (239).

Estlander went through several phases on the language question, beginning in a strong Fennophile position, shifting gradually into liberalism, and ending in a clear Svetophile position. Even early on, however, his Fennophilia differed sharply from the national idealism of the Fennomania stance, as was clear in his 1856 article in the *Helsingfors Tidningar*. According to Estlander Finland's bilingualism was not the problem, nor even the cause of the problem; rather, it was the effect or symptom of something else, namely a class conflict. Swedish was the language of the upper classes, Finnish of the lower. Once the real problem had been identified, he argued, one could begin to redress it in earnest: not by polemicizing for one language over the other but by breaking down the social barriers between them; by rapidly increasing education in Finnish, for example.

An important turning point in Estlander's thinking on the language question came in 1869, when he reviewed Yrjö Koskinen's Finnish-language textbook, *Oppikirja Suomen kansan historiassa* ("Textbook on the History of the Finnish People"). There Estlander took issue with Koskinen's separatist treatment of Finnish political history during Swedish rule, as if Finland had not then actually been part of Sweden. The established position was that Finland's "own" history under Sweden was cultural history; Koskinen, like many others of the day, was attempting to expand Finland's national uniqueness to cover political history as well. This review severed Estlander's ties with the Fennomania movement; and he later went on to insist that Finnish culture had long been, and would continue to be, supported by the Swedish-speaking intelligentsia. He was adamantly opposed to the one-to-one equation of nationality with language. As the century progressed the language question became increasingly more heated, to the point where, in an 1880 letter to a Swedish colleague named C. R. Nyblom, in connection with a controversy over an appointment, Estlander characterizes the Fennomaniacs in the Division of History and Linguistics as "that American gang."[81]

Estlander's own first scholarly work, like that of Rancken, Collan, and Toppelius, reflects the centrality of the *Kalevala* and folklore studies in the middle of the nineteenth century. He is also drawn to questions of nationality, particularly literary phenomena that adumbrate the struggle of an oppressed people against their oppressor.

In his 107-page dissertation on Richard the Lionheart in history and poetry (1858), he uses the *Coeur-de-Lion* text to paint a picture of medieval literature. He approaches his subject in two parts, the first dealing with Richard the Lionheart in history, the second with the origins of the medieval chivalric romance, in particular with two surviving English translations of *Coeur-de-*

Carl Gustaf Estlander (SLF)

Lion. He takes his historical background largely from the work of Augustin Thierry, but his discussion of the English Lionheart texts is based on close reading in sociohistorical context. Just as the emerging institution of knighthood was increasingly cut off from national and social interests, Estlander argues, so too does folk poetry begin to lose its national character as it is transmogrified into chivalric poetry. In folk poetry the unique voice and character of a nation could be heard; characters acted against a backdrop of national thought and customs, and the poet's perspective on the action was firmly grounded in that backdrop. As chivalry became increasingly dominant, the national or "folk" features of poetry were seen as merely provincial and displaced by cosmopolitan features. In this process the popular (*folkliga*) customs and values that had once grounded poetic heroes' actions begin to make room for the individualistic values of the knights: honor, nobility of spirit, gratitude, love, melancholy, jealousy, etc. Even so, however, the poetic presentation of these values remained steeped in the mentality of the times. On the one hand the narrator and his audience *believed* in the basic heroic situation of the tale; this is clearly evident in the frequent confusion between fiction and nonfiction, poetry and history, in the period. Chivalry offered a powerful explanatory model for poets and historians alike. On the other hand, the narration developed a particular stereotyped style – Estlander stresses the paucity of nature description in these tales, for example – and poets chose as their narrators the professional craftsmen of that style, singers, whose various types Estlander discusses at length.

The 83-page dissertation Estlander wrote for a docentship, *Folksångerna om Robin Hood* ("Folk Songs About Robin Hood," 1859), is closely related to the Lionheart study, though it is more emphatically historical in approach. He argues that "a folk song cannot be understood unless it is understood as an evolutionary process" (10). He traces the evolution of various songs against their historical backgrounds, like Sir Walter Scott and Augustin Thierry dating the Robin Hood songs in the twelfth century and seeing them as Anglo-Saxon expressions of resistance to the dominant Normans. Estlander attempts to trace the historical evolution of the songs, showing how in the early songs Robin Hood is portrayed as a noble, respected leader, but by the later songs has become a hardened bandit who robs the poor as well as the rich. By using this method Estlander seeks both to flesh out a picture of historical reality and to reach through the songs to the "original image" or *Urbild* they reflect and refract.

During his studies, and in the research that developed out of those studies, Estlander formed a conception of literary criticism as fundamentally a historical discipline. This placed a certain methodological distance between Estlander and Cygnaeus, which Estlander himself described in several places. In his 1893 review of Cygnaeus's collected works, he notes that Cygnaeus did not write a single study of literary history; Cygnaeus was instead a straightforward literary critic whose main concern was to determine a work's or a literary period's moral or aesthetic value. "For in order to trace an historical development one must discern a law, idea, hypothesis, or other concept that unifies change in time or space – and Cygnaeus had no inclination toward this sort of research" (373). Estlander's own historical, or, as Olof Mustelin calls it in his biography of Estlander, "historico-sociological," approach is most evident in his early scholarly work.

In the fall of 1858 Estlander and Toppelius applied for the same travel grant from the university senate to do folklore research in Central Europe; both men argued in their applications that European folklore scholarship would greatly benefit Finnish work in the field. Neither got the grant; it went to Edvard Grönblad, a historian. The senate debate on the various candidates is interesting for the light it can shed on the status of literary criticism in the period. Wilhelm Lagus, professor of Oriental literature, wrote in his voting statement that "aesthetic and literary-critical knowledge can be of no help in understanding the national epic," because "it is characteristic of poetic works of this sort that they are totally independent of the knowledge in question."[82] At just about this time, folklore was already in the process of being subsumed into Cygnaeus's discipline; Lagus's opinion is significant in its anticipation of later developments, which split folklore off from both literature and Finnish language and made it a separate discipline.

In any case, Estlander was well-to-do, and thus not dependent on travel grants; in November, 1859 he set off for Central Europe, where he visited art museums, did archival research, and worked in libraries in Berlin, Dresden, Leipzig, Munich, and Paris. In Berlin he met Wilhelm Schott and Jakob Grimm, in Dresden Hermann Hettner; but the young Estlander did not find these meetings as rewarding as he had expected. Hettner, that "untrammeled, independent thinker" and "gifted stylist" whom he had looked forward to meeting, was in fact a "grouchy tiresome man who, with his bad coloring and slouching posture, resembled nothing so much as a frog" (*Ungdomsminnen* ["Recollections of Youth"] 325-334). His acquaintance in Paris with Paulin Paris proved more fruitful; Paris directed his interests toward medieval literature in the Romance languages.[83]

When Estlander returned to Finland in August, 1860, he assumed his newly-appointed position as docent in aesthetics and modern literature – the appointment had gone through while he was in Central Europe. In the autumn term he lectured, as he had promised Cygnaeus he would in a letter written in April, "on modern epics beginning with the *Poema del Cid* and ending with the Nordic Eddas" (quoted by Schybergson, 89). In the same letter he admitted that he would rather not lecture on aesthetics, because many of the central issues in that field were still unfamiliar to him.

In the 1861-1862 academic year, the year after his appointment to the docentship, he deputized in Cygnaeus's professorship, and continued his lectures on modern epical "national poetry." He also lectured on a slightly more aesthetically oriented topic, "fantasy and artistic production"; and the following year, back in his docentship, he gave a lecture course in aesthetics (on topics requested by his audience) and another in the history of painting. Estlander was gradually branching out into the various sectors of the discipline. His next extensive trip to Europe was planned, in fact, around a projected study of painting; he received the large Alexander stipend and headed south in the fall of 1863.

Before he left, however, he managed to publish some of the results of his previous trip: the Finnish Society of Sciences and Letters brought out his *Poema del Cid i svensk öfversättning med historisk och kritisk inledning* ("The Poema del Cid in Swedish Translation, With a Historical and Critical Introduction," 1863), which included his own Swedish translation of the Cid poem and a 78-page commentary covering the secondary literature, the hero's historical model, the origins and structure of the poem, and his translatorial principles. In his translation Estlander drew heavily on Damas Hinard's French version of the poem and Dozy's Dutch study of it. In his commentary he says that he set himself the task of determining both the individual and the collective elements of the poem's origins. By "individual" elements he means the composition and structure of the poem at the hand of an "old monastery brother" in the twelfth century: Estlander speculates on the compiler's identity based on internal evidence, particularly the frequency of Latinisms and religious phrases. By "collective" elements he means "portraits of character and customs," which must be regarded, he insists, as "original national songs." Since the compiler did not reshape these latter, Estlander argues that the poem as a whole must be considered an ancient and authentic example of Spanish national poetry.

During Estlander's two-year journey through Europe he familiarized himself with art in Stockholm, Copenhagen, Paris, London, Florence, Rome, and Madrid. Simultaneously he assiduously collected materials for a study of medieval literature in the Romance languages, particularly the Tristan story and Provencal literature. In Provence Estlander met the 25-year-old leader of the Provencal cultural organization called Félibres, Frederic Mistral, who later won the Nobel Prize for Literature. Estlander tried in vain to instil in Mistral the national-idealist Finnish model of linguistic identity. For Mistral it was enough that Provencal be used and developed in poetry alone; it should not be made the language of education and government.[84]

Estlander's interest in Provencal literature was, of course, nationalistically motivated. In the 1905 article he wrote on the occasion of Mistral's nomination as the 1904 Nobel Laureate in Literature, he says that Provencal culture interested him originally because in Provence, as in Finland, there was an opposition between the language of the people and the language of the cultured elite.

As on his first trip to Europe, Estlander met many other well-known scholars as well; in Paris, for example, he met the respected Orientalist Garcin de Tassy, and helped to found the Société de linguistique in 1864. Estlander returned to Finland in the fall of 1865.

During the next three years Estlander was a prolific scholar. Based on what he had seen and experienced in art museums and archives all across Central Europe, he lectured in the spring of 1866 to an exceptionally large audience on the history of modern art and wrote *De bildande konsternas historia* ("The History of Pictorial Art," 1867), a history of European (and particularly Nordic) art from the late eighteenth century to the present. The published work brought him a good deal of prestige and acclaim throughout the Nordic countries.[85]

In this period preceding Cygnaeus's retirement, Estlander published two other books as well, both on medieval European literature. One had been inspired by Paulin Paris, "Pièces inédites du roman de Tristan, précédées de recherches sur son origine et son développement" (1866); it was a Swedish translation of Tristan with an introduction in French (the only work Estlander published in a major language) explaining the versions used and tracing the history of the text.[86]

The other, *Bidrag till den provencaliska litteraturens historia* ("A Contribution to the History of Provencal Literature," 1868), was Estlander's professorial dissertation, a 146-page thesis that he defended in May, 1868. In his introduction, Estlander explores the beginnings of scholarly interest in Provence in the early nineteenth century and the increasing fascination it has held for scholars to date, and hopes that Provencalism will expand into the social and political realms as well as the poetic. In his literary-historical section he stresses the importance of examining not only the Golden Age of Provencal literature, the age of the troubadours, but also that which followed it, the age of language and literary history, which clearly shows how poetry "under the same sky and at the service of the same national character changes under pressure from changing social and political circumstances." Estlander planned to trace the history of Provencal literature into the nineteenth century, but in his dissertation only reaches the sixteenth century. It is characteristic of Estlander's socio-historical method that he attempts to determine the social causes of the decline of Provencal literature and possible causes of a new rise to glory.

In the space of ten years Estlander had published five literary-critical monographs – more than any other Finnish literary critic before him – and having deputized for Cygnaeus in 1861-1862 and 1867, he was a strong candidate for the professorship when Cygnaeus retired in 1867. He had also served as official faculty opponent in three dissertation defenses, Edman's, Collan's, and Toppelius's. By the application deadline in 1867, only Estlander and Toppelius had applied for the post. In his application Estlander requested, in light of his scholarly credentials, that the required professorial dissertation in his case be waived.[87] The university senate asked the Division for History and Linguistics to evaluate Estlander's scholarly qualifications to date, and in a lengthy statement docent in philosophy Wilhelm Bolin did so. The Cid study, he wrote, demonstrated a strong familiarity with previous Cid research and modern research methods. The Tristan commentary too was laudable in its familiarity with medieval literature and "sharp and formidable" use of scholarly argumentation. Bolin also noted that the Tristan commentary operated in "the most important [förnämst] area of contemporary literary criticism," a comparative approach that was of the utmost importance for Finnish national literature as well. Bolin praised the history of European art for its excellent disquisitions on the realization of aesthetic principles in modern art and on the relationship between art and reality. In his conclusion Bolin states that Estlander's three scholarly works demonstrate a comprehensive erudition both in aesthetics and in modern literature. The division approved and accepted Bolin's statement on November 8, 1867.

The senate debate on Estlander's petition to waive the professorial dissertation was divided. A majority voted against it, arguing that despite Estlander's incontrovertible scholarly credentials it would be unfair to the other applicant to waive the dissertation requirement. It was further surmised that comparison between the two candidates might be difficult if only one had written a professorial dissertation. In his dissenting opinion, professor of Roman literature Edward af Brunér questioned the value of the professorial dissertation in general.

Having denied Estlander's petition, the university senate granted both candidates a respite period until October, 1868. Estlander finished his professorial dissertation in May. Collan was appointed official faculty opponent, and while he found the dissertation generally meritorious, he did have one major criticism: Estlander emphasized the political function of the troubadour poetry, he said, at the expense of its erotic function. The attention Estlander paid to the later developments of Provencal literature Collan found particularly noteworthy, since there was no precedent in the secondary literature for this sort of study.[88]

Meanwhile, Toppelius had made no progress on his own professorial dissertation, and accordingly had withdrawn his application. On June 3, 1868, therefore, the university senate recommended that Estlander be appointed to the professorship, and he received the appointment in July.

Estlander's scholarly career can be divided into three distinct periods. The first focused on the study of medieval literature, the second on art history and practical cultural politics, and the third on the study of Runeberg. Before his appointment to the professorship his work had revolved around the earliest literature of Central and Southern Europe; after his appointment he showed no further interest in that subject, until his retirement, when he published a collection of his earlier lectures, *Vitterhetens utveckling hos de nyare folken. 1. Medeltiden. Förra perioden* ("Literary Development Among the Modern Peoples. 1. Middle Ages. Earlier Period," 1900). Provencal literature played a central role in this work, which, as the volume number in the title suggests, was intended to form part of an extensive history of world literature.

In the late 1860s, with the appointment to the professorship, Estlander's career shifted increasingly away from scholarship and into the practical arena of cultural politics. He was instrumental in the founding of a new Swedish-language cultural magazine, *Finsk Tidskrift* (1876-), the Svenska Litteratursällskapet i Finland (Swedish Literature Society in Finland, 1885), and the Ateneum Art Museum (1887), and played an active role in the management of these enterprises.

His political shift on the language question is reflected in the editorial policy piece he wrote as editor-in-chief of the *Finsk Tidskrift* in 1876, "Till läsaren" ("To the Reader"). There he speaks of the dual roots of Finnish literary culture, the bilingual national literature that records half a millennium of cultural endeavor. To cling to only one branch of this cultural heritage, that written in Finnish, would be as absurd as to reject half of one's parental heritage on the grounds that "it is half aristocratic, half peasant stock."[89] Estlander's dream was to make the journal a semischolarly forum for all ideological positions that would not be mired in the muck of ordinary politics. But this was a pipe-dream in the political climate of the 1870s; the Fennophiles rejected the magazine, as Eliel Aspelin's refusal to collaborate made clear.[90] Behind the increased polarization of the two language camps in the 1880s lay the fact that a whole generation had been raised and educated in the spirit of Finnish nationalism, and had gradually moved into positions of power in the university, in schools, in the government bureaucracy, in the courts, in literature, and in the press. Simultaneously the mediatory liberal party had withered away, leaving an open field for the emergent Svetophile party.[91]

While the *Finsk Tidskrift* was specifically a general-cultural publication, it proved influential for literary criticism, as its book review department kept alive the *Litteraturblad* tradition of introducing the latest literary-critical developments, particularly in the Nordic countries. In addition to Finnish critical works, new scholarly works by Warburg (1876, 1881, 1884, 1890), Brandes (1878, 1894), Ljunggren (1882, 1887), Schück (1886, 1888, 1889, 1892, 1895), and Levertin (1890) were reviewed.

Even more significant for the development of Finnish literary criticism was the Swedish Literature Society in Finland, which published an enormous amount of research on Swedish-language literature and the history of learning, literary-historical documents, and bibliographies in its publication series. One of these was a ten-volume history of the history of learning at the University of Turku, *Åbo universitets lärdomshistoria* (1890-1908), which Estlander proposed to M. G. Schybergson in a letter in 1885, and was approved by the society in 1888 (Steinby 55).

Research on art history in Finland in the mid-nineteenth century required extensive travel abroad, and Estlander left on his first long trip abroad (for almost a year) in July 1873, with the idea to study the European art industry and vocational training. He did very little literary-critical research on this trip, but in his travel memoir, *Vid konstflitens härdar i Tyskland, Österrike, Schweiz och Belgien. Reseanteckningar* ("In the Centers of the Art Industry in Germany, Austria, Switzerland, and Belgium: Travel Notes," 1875), he relates his meeting with Friedrich Vischer in Stuttgart: "When one is aware of the

The university hall. A. Federley's drawing, 1880s.

implacability with which Vischer has defended himself in endless battles with critics and parasites, one cannot help but be surprised to view the modest, almost apologetic man before one, who seemed to differ from other great men in another respect as well, that he listened as willingly as he talked." Despite his reservations, it is clear from this account that Estlander still greatly admired Vischer's aesthetic system and the man who, having created it in his younger years, still kept revising it in his old age.[92]

Estlander's fourth extended trip to Europe, from August of 1883 to January of 1884, was similarly dedicated to art-historical research. On this trip he visited Lübeck, Dresden, Florence, Rome, and Sicily, working on a book on Hippolyte Flandrin that was not published until 1890.

Beginning in the late 1880s Estlander began to move back into literary-critical research, with a central focus on the work of J. L. Runeberg. Runeberg was now doubly suitable to Estlander's goals: he was the major Swedish-language writer in Finland, who thus represented the Swedish-speaking cultural heritage for whose survival Estlander felt it so important to fight; and his lucid, idealistic world view appealed powerfully to Estlander's tastes, which the new literary trends of the 1880s did not satisfy. His view of literature in this period is best reflected in two articles, "Det estetiska och moraliska intresset i skaldekonsten" ("The Aesthetic and Moral Interest in Poetic Art," 1882), which he published in the *Finsk Tidskrift*, and "Naturalismen enligt Zola" ("Naturalism According to Zola," 1891), which appeared in an invitation to a professorial inauguration. In both pieces Estlander stresses the importance of aesthetic harmony, and distances himself from naturalism. In the former article he writes, perhaps rather pathetically: "We are so old-fashioned as to be unable to find beauty in a work that leaves us frustrated before an unsolved or even unsolvable conflict." Such a work, according to Estlander, was Ibsen's *Ghosts*, published the year before, which lacked "that reconciliatory and reassuring ending that is the distinguishing characteristic of tragic poetry." A true writer, Estlander claims, fills his work with harmony: "he soothes us with the certainty that good will still win out over bad, that justice will banish injustice, and if something incomplete or unpleasing remains in this life, he points ahead to the life to come."

In an early (1857) *Litteraturblad* article on Runeberg's "Svartsjukans nätter" ("Nights of Jealousy"), Estlander offers a symptomatic definition of Runeberg's special genius. In both Runeberg's epic and lyric poetry, there is a harmonious balance between the material and the spiritual in which the material is given its full due and the spiritual is portrayed through its appearance in the material. The erotic "Svartsjukans nätter" does not entirely reach this ideal, and Estlander's lengthy article sounds rather like an attempt to justify Runeberg's poetic practice.

In the late 1880s there was no need to write a biography of Runeberg, since J. E. Strömborg was in the process of publishing a multivolume collection of biographical notes on him. Instead Estlander decided to write on Runeberg's world view, a decision that, as Estlander's biographer Schybergson writes, had a deeply personal element: Estlander believed that Runeberg offered the best possible model for human existence, and through his study sought confirmation for his own beliefs (510). There is, however, another and more practical aspect to Estlander's personal motivations in a paper on Runeberg's religious world view that he gave at the Swedish Literature Society's first annual meeting. Part of it, certainly, is personal: having set up King Fjalar, who first defies fate and then humbles himself before it, as the ideological key to Runeberg's work, Estlander claims that "our age seeks blustery and selfish glory like Fjalar," but in the end it too must be humbled. But another part is political, tied up in the practical realities of advancing Swedish-language literature: the Swedish Literature Society in Finland had been founded in part to honor and advance the memory and heritage of Runeberg, and Estlander wanted to instigate new Runeberg scholarship, which had come to an almost complete halt after Cygnaeus. Estlander insisted especially on the importance of *Finnish* Runeberg scholarship, since he felt it was difficult for a Swede to appreciate the Finnish complexity of Runeberg's world view – the influence of Pietism, for example, a point he had drawn attention to in an 1871 review in the *Helsingfors Dagblad* of the Swedish scholar C. R. Nyblom's edition of Runeberg's collected works.

J. L. Runebergs estetiska åsigter ("Runeberg's Aesthetic Views," 1888) is an expanded, 88-page version of that Literature Society paper that deals specifically with the aesthetic side of Runeberg's literary theories. Drawing on Runeberg's academic dissertations and later articles, reviews, and letters, Estlander attempts to determine how Runeberg saw the relations between art and religion, art and life, ancient and modern, and how Runeberg understood such concepts as "idealism," "idea," "ideal," "idealization," "naturalism," "realism," "fantasy," "ugly," "comic," and "sublime." In Estlander's view Runeberg's ideal is a "realism" that is based solidly in reality, but idealizes it, frees its depiction of all inessential elements of reality. In Runeberg's theoretical pronouncements idealization moves consistently from the whole to its parts, which means that the poet must carefully weigh the relation of the details he portrays to the "wholes" of law, idea, or nature.

Estlander's study is based on a close philological analysis of Runeberg's texts, and he is very aware of the many contradictions in them; he also sets up an evolutionary or genealogical model of the development of Runeberg's views beginning in the early nineteenth century. Despite this historical-philological method, however, Estlander's approach to Runeberg is static. He does not in the end see Runeberg's views as a historical process, but rather as the end result of a coherent personality. Thus he concludes by stressing the hegemony of ideas, "the great advantage of authentic art over what we see around us: when unity at last opens clearly to our eye, it has a rebounding effect back on the details and the process of which it is itself the result, and thereby confirms their necessity." The reference to "authentic art" and the hope that "Runeberg's views of art might still hold sway over contemporary opinions," like the remarks he made elsewhere on the inadequacy of realistic and naturalistic art, are utterly typical of Estlander's late scholarly vision.

In the early 1890s there was talk of bringing out a Finnish "normal" or critical edition of Runeberg's works. By the time this talk became a reality, towards the end of the decade, Estlander had already retired, but he accepted the editorship of the eight-volume work nonetheless. As it happened, his assistant Hjalmar Appelqvist did most of the practical editing work for him, though Estlander decided on certain key principles and wrote a 300-page commentary entitled *Runebergs skaldskap* ("Runeberg's Poetry," 1902) that was appended to the series. This was the first comprehensive Finnish study of Runeberg, who by this time had been canonized as Finland's "national writer." Estlander provides considerable biographical information on Runeberg, which he largely adapted from Strömborg's notes, but the interpretation of texts takes clear precedence over biography in the study. Estlander also draws on other recent Runeberg criticism and makes excellent use of early drafts of Runeberg's works, and his discussion of the composition of the *Elk Hunters* is far superior to the illustrious study published in the 1880s by the Swedish scholar Ljunggren.

Behind the heavy emphasis Estlander places on Runeberg's Swedish-speaking family background and education there seems to run something like the following train of thought: Runeberg stands apart from both Swedish-speaking writers from Sweden and Finnish-speaking writers from Finland, in that his work is grounded in both the Swedish-language culture of Finland and a deep knowledge and understanding of Finland's Finnish-speaking people. Estlander does not stress the national aspect of Runeberg's work, as Vasenius was to do soon after; Estlander's Runeberg gains his stature as the poet of a religious world view, right thinking, and humanity. Indeed, in comparison with Werner Söderhjelm's biography of Runeberg (1904, 1906), Estlander's study is clearly the work of an old-school scholar. As in his earlier studies of Runeberg, Estlander stresses the poet's "realism," which conveys to the reader a deeper, more idealistic picture of life than the harsh surfaces cultivated by the naturalists.

Estlander's professorial lectures fell into three categories. The central one of these was literary history, most importantly medieval literary history. He lectured (first as a docent, later as the professor) on the epic folk poetry of the modern peoples (1861-1862), the history of French literature (1868-1869, 1885-1886, 1890-1891), the history of Southern-French, Italian, Spanish, and Portugese literature (1869-1870), the history of English literature (1870-1871, 1887-1888, 1888-1889), the history of German literature (1871-1872), the history of Swedish literature (1872-1873), the history of Renaissance literature (1875-1876, 1876-1877), the history of eighteenth-century literature (1876-1877, 1877-1878, 1881-1882, 1882-1883), the history of eighteenth-century Scandinavian literature (1878-1879), the history of medieval literature (1879-1880, 1891-1892, 1896-1897, 1897-1898, 1898-1899), the history of modern literature in the Middle Ages and Renaissance (1880-1881), the history of Provencal and Italian literature (1884-1885), the history of Swedish literature beginning with Dalin (1887-1888), the history of nineteenth-century English literature (1889-1890), and the history of modern literature (1892-1893, 1893-1894, 1894-1895).

A second important teaching area was aesthetics and literary theory. Estlander gave an introduction to aesthetics course numerous times during his career (1862, 1869-1870, 1872-1873, 1875-1876, 1892-1893, 1896-1897), and another, with numerous variations, on poetic theory (1861-1862, 1870-1871, 1876-1877, 1878-1879, 1885-1886, 1888-1889, 1889-1890, 1893-1894, 1897-1898).

His third teaching area was art history. He gave two lecture courses on general art history (1863, 1866-1867), one on art history in the Italian Renaissance (1871-1872), one on art theory (1879-1880), and three courses combining theory and history (1881-1882, 1882-1883, 1883-1884). Estlander's teaching in art history came to an end in 1884, with the appointment of J. J. Tikkanen (1857-1930) as docent in aesthetics and art history.

Estlander retired from the professorship in August, 1898. The same year he was awarded the honorary title of Councilor of State, and was raised to the peerage, both awards in recognition of his contributions to Finnish culture. Estlander's professorial work was not without its weaknesses; the number of actual dissertations written, for example, did not begin to rise until toward the end of his tenure, and his professorship was never particularly fruitful for Finnish-language literary criticism. Still, Estlander's long tenure as professor laid a foundation for increasingly international scholarship among the younger generation of literary critics; while he was not an international scholar in the sense of publishing prolifically outside Finland, in foreign languages (almost everything he published was in Swedish), he made many contacts with prestigious scholars in European universities, especially in Sweden. Beginning in the 1870s Estlander was honored with several academic awards for the active part he played in bringing Finnish and Swedish scholars together. C. R. Nyblom, professor of literature at the University of Uppsala, was one of Estlander's long-term friends, and their extensive correspondence provided Estlander's biographer Schybergson with one of his most important sources of information. Both Nyblom and the young August Strindberg wrote briefly for the *Finsk Tidskrift*. Indeed, in the early 1880s Estlander was offered a teaching job at the College of Stockholm (Stockholms Högskola) – a fact Schybergson never mentions. In a letter to Snoilsky on November 12, 1883,

Estlander explains why he cannot accept the offer: he is still needed in Finland, he writes, to help strengthen the position of Swedish-language culture.

Kaarlo Bergbom and the Suppression of the Rising Generation

The first dissertation to be written under Estlander was defended in the spring of 1868 by the 24-year-old Karl Johan Bergbom (1843-1906). His 101-page study of the German historical drama, *Det historiska dramat i Tyskland* ("The Historical Drama in Germany"), was the first dissertation written in Finland on modern German literature.

Bergbom was born in Viipuri, son of a Fennophile bureaucrat who helped found the Viipuri Finnish Literature Society in 1845 and was elected to the Senate in Helsinki in 1850.[93] Kaarlo Bergbom – he Finnicized his own Christian name, and it is by that name that he is known – picked up his father's Fennomania and studied Finnish assiduously, even to the point of living in rural Finnish-speaking areas for long periods. He was also involved in several Fennomanic cultural projects: along with J. V. Calamnius, Jaakko Forsman, and Frithiof Perander, he helped found the influential *Kirjallinen Kuukauslehti*, and, most important of all, he was instrumental in the founding in the early 1870s of the Finnish Theater (now the National Theater) in Helsinki. He then served as the Finnish Theater's first director, a post he held until the early years of the new century.

At the university Bergbom had studied aesthetics and modern literature, history, philosophy, and Finnish. In the spring of 1863, when he completed his Master's degree at the young age of 19, his plans for the future were still somewhat hazy. At first he thought he might become a writer in Finnish, and indeed wrote several short stories and, later, book reviews in Finnish. But his literary efforts, including a play in Swedish, are not particularly impressive, and his Finnish was never very fluent.

His second dream was to become a scholar. In the fall of 1864 he gave a long lecture at the Porthan celebration held by the Ostrobothnian Students' Club on Heinrich Heine, and the paper, which was not published until Bergbom's collected works were brought out in Finnish (1907, 1908), demonstrates a commendable familiarity with contemporary German literature. It also reveals a Hegelian-Snellmanian conception of literature as a mirror of the *Volksgeist*:

> The spirit of every nation gives birth to a few such individuals and breathes into them all that force that it has been given to do great deeds, all the hope of which it has been allowed to dream. (137)

According to Bergbom, the romantic Heine was not one of these true geniuses, these writers born of the *Volksgeist*; he merely reflects "human existence at specific points in time" and expresses "the hopes and the fears of the moment." Still, this too can be significant, particularly at great crises in history and in the "great cultures of Europe."

In his study of Heine Bergbom conceives of the development of literature and culture dialectically: the one-dimensional cultural Germany of Goethe had emptied out into the one-dimensionality of political Germany, and Heine reflects this historical shift negatively, as a critic, satirist, ironist. Somewhat in the same mode as Cygnaeus's study of the *Songs of Ensign Stal*, Bergbom sees Heine's satire as a negative phenomenon rising out of Jewish bitterness, which must soon vanish in the face of a new synthesis of culture and politics.[94] Bergbom also draws certain correlations for Finland:

> Although the obstacles that seem to block our growth on a national basis are not identical to those in Germany, they are still similar enough for us to learn valuable lessons from them – if not so much what we must do as what we must *not* do. The four parallel trends that I have outlined as characteristic of German life in this century are to be discerned in our country as well. And just as we confidently believe in the truly Finnish Finland which the future shall create for us, both in language and in culture, so would we hope the ever growing national spirit will unite this great German culture into the single whole they seek – hopeless as this prospect may look on several fronts at the present time. (170-171)

Beginning in 1866 Bergbom wrote book reviews for the *Kirjallinen Kuukauslehti* for half a decade, primarily on Finnish-language literature. At the same time he continued to work on German literature for his Licentiate of Philosophy degree. In *Det historiska dramat i Tyskland*, he first reviews the status of the historical drama in world literature and the history of German drama since the Middle Ages. The main period of the historical drama, beginning in the eighteenth century, he divides into three periods: the humanistic, the romantic, and the Young Germany period. Bergbom's secondary sources include the literary histories of Robert Prutz, Rudolf von Gottschall, Julian Schmidt, and G. G. Gervinus, but he primarily uses them to orient himself in his subject. His interpretations of specific plays are original, based heavily – especially when he considers their performability – on his practical experience in the theater.[95]

In connection with Schiller's plays Bergbom places particular stress on "general historical ideals" and "historical forces." In *Wallenstein*, for example, "general historical ideals have taken the shape of fate, a force that remains beyond the ken of the characters in the play, who merely follow the lead of their own subjective passions, but that is nevertheless evident to the viewer, since world history appears in it in the form of inevitability" (44). In *Wilhelm Tell* and the *Jungfrau von Orleans* "national habit [creates] the pathos of freedom: the general determinant forces have been violated, and the greatness of the heroes lies not in the arrogant bravado with which they sever their ties with historical forces, but rather in their willingness to be possessed mind, body, and soul by those forces and to give themselves over to their direction" (54).

On the whole Bergbom's attitude toward romanticism is negative, even though he admits its significance in the birth of nationalism. He attacks the works of the romantics not only as a pragmatically minded theater director but also as a Protestant: he finds "mysticism" and "Catholic unfreedom" distasteful. In the Young German playwrights Bergbom discerns an inner conflict, though he praises Grabbe, for example, "the Feuerbach of drama," for his masterful handling of crowds. Bergbom's critiques of German plays, the most recent of which was written in 1867, are sharply distinctive, filled with the hands-on approach of the skilled director. Indeed, in his analyses Bergbom consistently stresses dramatic structure and characterization at the expense of philological or historical features.

Bergbom's dissertation came at a critical phase in the development of the discipline. Cygnaeus, who returned from retirement to chair the defense, said in his opening remarks that he had returned to the podium "like a farmer returning to the furrows in which he had sown his finest seeds, and which in their fullest bloom hold promise of rich, ripe harvests."[96] In Bergbom's case the prediction was premature. Bergbom was a programmatic Fennophile, and the new professor was already shifting in a distinctly Svetophile direction. The first sign of the schism that was already then taking shape was Bergbom's review of Estlander's book on art history in early 1869, where, in the midst of praise, Bergbom criticized Estlander in lofty Snellmanian terms for neglecting nationality. Estlander did not sufficiently make clear, he said, when art in a given country had been "the necessary product of the national spirit, the perfect expression and depiction of a certain secret life-giving force." As we have seen, this sort of talk was entirely at odds with Estlander's approach to literature, and that, combined with Bergbom's Fennomania and inclination toward Germanic rather than Romance and Anglo-Saxon literature, made it difficult for Bergbom to advance his academic career.

After defending his dissertation, Bergbom was planning to travel to Europe to collect material for his docent dissertation on European literature and art, but he did not receive the Alexander stipend he applied for, and had to postpone his trip until 1871, and then to travel at his own expense. He traveled via St. Petersburg to Berlin, where he spent most of his time in the theater. He also spent some time in the library, however, planning a lengthy article on German literature from 1848-1870, to be published first in the *Kirjallinen Kuukauslehti*, then, in an expanded form, as a separate book. This plan was never carried out. From Berlin he submitted a second application for a travel grant, this time in order to study the literature and art of small countries in which two cultures coexisted side by side – an indigenous culture and a foreign culture brought in by historical circumstance – and which found themselves in a transitional state in the nineteenth century. This was certainly an interesting proposal, but it was tarnished with the brush of the "language question" in the university senate, and Estlander refused to support his student's application. Professor of history Yrjö Koskinen gave Bergbom his support in the face of Estlander's opposition, but it was not enough. The grant application was denied.[97]

Bergbom was bitterly disappointed by this decision, and his disappointment seems to have been shared by the other Fennomania people. So at least it appears from Snellman's remarks in the summer of 1871: he sent Bergbom a letter via Bergbom's sister, advising him to give up his studies in aesthetics and modern languages. "It will never wash in this country," he wrote, "never wash, and it won't put bread on the table."[98]

Even in the late 1870s, however, Bergbom was still planning a history of world literature, to be published by the Finnish Literature Society at a length of around 1000 pages. The society specified that it be a proper literary history

For decades the Ekberg café on Aleksanterinkatu was a favorite gathering spot for students. A. Federley's drawing from the 1880s.

written in "solid scholarly prose," rather than a glorified bibliography or textbook.[99] In 1879, while familiarizing himself with theatrical life in Budapest, Bergbom sketched out his plan for the work as follows:

> I've . . . been thinking of starting with a few sections – the Jews, the Romans, and the Renaissance – in order to show you in practice how I approach my subject, and then let you decide whether I should proceed with the remaining sections along the same lines. The Oriental literature and the medieval will work me the hardest, on the one hand because in those areas I am familiar with the fewest works, on the other because historical research is progressing with such gigantic leaps and bounds that what one read a few years ago with great satisfaction today looks out-of-date.[100]

Bergbom's own conception of the magnitude of the project was much greater than the society's; he predicted that the work would "exceed the society's bounds many times over."[101] In actual fact, by this time Bergbom seems to have devoted himself body and soul to the Finnish Theater, which under his direction was to become an influential institution for Finnish-speaking culture.[102]

Bergbom's dissertation was the last to be defended under Estlander for over a decade. It was not until around 1880 that a new generation of scholars emerged: Vasenius, Aspelin, the aesthetician Edvard Rindell (1858-1928), and Söderhjelm. The lack of doctoral dissertations in the discipline also meant a lack of teachers: after Toppelius stopped lecturing in 1869, Estlander was the only instructor in the field for the next decade, and when he was on leave of absence traveling abroad in 1873-1874, no lectures in the field were held at all. New life was not brought into the discipline until the turn of the century, through a methodological breakthrough with which Estlander had very little to do.

Early Kalevala Scholarship

It was to be expected that, upon its appearance in the 1830s and 1840s, the *Kalevala* would become the center of a great deal of both scholarly and journalistic attention. Public discussion commenced immediately upon Lönnrot's publication of the first edition in 1835, and continued with undiminished intensity when he brought out the second, expanded edition in 1849. The *Kalevala* was the first indication to the Finnish cultural elite that significant literature might be written in Finnish; indeed, since the *Kalevala* was seen as a collectively created work of art, the romantic notion of the people or "folk" as the true subject of artistic creation favored it over the more recent self-conscious or, in Schiller's term, "sentimental" literature.

It would in fact be four decades before the romantic epic-theory would finally disintegrate: until the 1870s, Lönnrot's epic and original folk poetry were simply equated. This equation is evident also in the fact that, with the notable exception of Cygnaeus's reading of the Kullervo story, the *Kalevala* was studied by scholars of the Finnish language rather than literary critics or aestheticians: it was not considered a work of literature or "art" at all. Lönnrot himself lectured on the *Kalevala* every semester beginning in 1854, and this tradition was continued by the later docents and professors of Finnish, August Ahlqvist and Julius Krohn, all treating the epic as the spontaneous creation of an entire people. Nevertheless, from a later standpoint (that of the 1880s and after), the work Finnish scholars had done on it was itself literary criticism – the *Kalevala* was, after all, unquestionably the literary creation of Elias Lönnrot.

The symbolic value of the *Kalevala* among the Fennomanic intelligentsia was enhanced by the fact that its folkloric material had been gathered in Eastern Finland and Russian Karelia, which had the effect of moving Finland's cultural center of gravity in the direction preferred by the Fennomania group: away from Sweden, away from the Southwestern coast of Finland dominated by the Swedish-speaking intelligentsia, toward the unwashed peasant heartlands. Suddenly, folklore-collection was all the rage among Fennomanic students.

It was not until the late 1870s that scholars systematically divorced Lönnrot's epic from folk poetry – and this divorce led to the birth of Finnish folklore studies, to the development of the famous "Finnish method" pioneered by Julius Krohn.[103] Another side effect of that divorce, however, was the scholarly neglect of Lönnrot's epic. The myth of the Finnish national epic was maintained for nationalistic and pedagogical purposes, but it was to be many decades, well into the twentieth century, after the Second World War (the Estonian scholar August Annist in 1944 and the Finnish scholar Väinö Kaukonen in 1979), in fact, before scholars would rediscover the *literary* text that Lönnrot created.

The earliest *Kalevala* scholarship, with its romantic conception of the epic as the collective creation of the Finnish people, portrayed Lönnrot as the mere restorer of a once-unified but now fragmented whole. It was in this spirit that J. G. Linsén, chairman of the Finnish Literature Society, spoke on the *Kalevala* before the first edition was even off the presses in 1835. Lönnrot, Linsén proclaimed, was the "discoverer and arranger" of the ancient epic, the patient editor who had taken great pains among the Karelians living in the District of Arkangel "to bring into full light of day what had crumbled into shards and fragments and fallen into neglect."[104]

The most influential romantic readings of the epic were published in the mid-1840s. Both J. V. Snellman and J. J. Tengström's son, philosophy docent Robert Tengström, applied Hegel's romantic epic-theory to the *Kalevala*, with the single difference that where Snellman was cautious in his claims, Tengström was exultant. According to Snellman, who wrote on the *Kalevala* in various places,[105] the Finnish people could not have given birth to a true epic, because Finnish history had never known a Heroic Age, which was required by the romantic epic-theory. Tengström, on the other hand, in an article entitled "Finska folket såsom det skildras i Kalevala" ("The Finnish People as Depicted in the *Kalevala*," 1845) originally presented at the annual meeting of the Ostrobothnian Students' Club in 1844, constructed a Heroic Age for Finland and thus paved the way for the later scholarly fashion of assigning the epic a historical foundation.[106] This romantic theorizing of the *Kalevala* was lent authority by the internationally renowned Jacob Grimm, who gave a talk at the Berlin Academy of Science in 1845 "Über das finnische Epos" ("On the Finnish Epic"). In that talk, which was published in the Helsinki-based *Fosterländskt Album* in Swedish translation later that same year, Grimm dated the origins of the epic back in the early mythic period, characterized it as a "pure epic," and regarded it as clearly superior to Ossian.[107]

Not even Lönnrot himself was particularly eager to analyze the role he had played in compiling the *Kalevala*, though in the process of editing the first edition back in the early 1830s he had recorded his realization that no unified ancient epic ever could have existed. Presumably the obvious parallel with Ossian, the late-eighteenth-century forgery by James Macpherson, made the question of who "wrote" the *Kalevala*, of its authenticity, a sticky one. Detailed knowledge of how the epic had come into being could, after all, be used as a polemical weapon. As early as 1839 A. I. Arwidsson had written a letter to Atterbom outlining a reasonably accurate appraisal of Lönnrot's actual role:

> Dr. E. Lönnrot has . . . collected a great number of old Finnish national songs or poems, some of which dealt with roughly the same topic or could at least be edited into a coherent whole. . . . He then gave this collection a name, despite its still rather loose structure and questionable inner integrity – the natural result of the way the whole came into existence. The Finns never had an ancient national epic named the *Kalevala*; rather, L. has attempted to gather up the old songs into a single whole and call the collection the *Kalevala*.[108]

Ten years later K. A. Gottlund would also, in his *Suomi* magazine, describe Lönnrot's role in the compilation of the *Kalevala* quite accurately – though largely in order to accuse Lönnrot of "corrupting" the folk poems by expanding and combining them into something "of the magnitude of Homer's *Iliad* or *Odyssey*" (23-24).

But these critical comments did not bring about a general awareness that the *Kalevala* was Lönnrot's creation. Even among scholars there was a strange "semiaesthetic" (in Rafael Koskimies's term) tendency to equate the *Kalevala* and folk poetry. There was very little "literary-critical" or "aesthetic" *Kalevala* scholarship at all in the first four decades of its existence, apart from sketchy debatės over whether its poems were mythical or historical, and whether it originated in Western or Eastern Finland. After Cygnaeus the most important work on the *Kalevala* was done by a docent (later professor) of pedagogics and didactics, J. J. F. Perander (1838-1885).[109] In his article "Tragillisesta peri-aatteesta Kullervo-runoissa" ("On the Tragic Principle in the Kullervo Poems," 1872), he argues that Kullervo is a tragic figure because fate has chosen him for vengeance and thus rendered him unfit for anything else, including life. Perander's aesthetic notions were not pathbreaking, but they played a significant role in the history of scholarship in that they were the first attempt to examine *Kalevala* characters and events on the basis of general theoretical concerns. Perander's articles on the tragic elements and female figures in the *Kalevala* throughout the 1870s and 1880s also inspired much turn-of-the-century art.

Even while Perander was giving lectures as a respected *Kalevala* scholar, however, a trend was already underway that would lead to increasing neglect of Lönnrot's text. The birth of Finnish folklore studies can be dated roughly between the years of 1872 and 1885. In the *Suomi Albumi* for 1872, Julius Krohn wrote an article on "Wirolaiset ja ylimalkaan Länsi-suomalaiset aineet Kalevalassa" ("Estonian and Western-Finnish Elements in the *Kalevala*"), arguing that the linguistic elements from Western dialects proved that the epic was not purely the creation of the Karelians but was "the precious shared heritage of the entire Finnish people from the era when tribal and dialectal differences were still taking shape." The next year, in an article entitled "Missä Kalevala on syntynyt?" ("Where Did the *Kalevala* Originate?" 1873), A. A. Borenius-Lähteenkorva (1846-1935) advanced historical, linguistic, and ethnological evidence to show that the ancient folk poetry had originated in Western Finland and then traveled to the areas in Karelia where it had been collected.

Theories of this sort compelled scholars to take a new look at the original folk poems. Before Julius Krohn the most significant pioneer of the new folklorist approach to the poems was Eliel Aspelin, in a single book entitled *Kalevalan tutkimuksia* ("*Kalevala* Studies", 1882). There Aspelin adapted E. B. Tylor's theory of animism in order to argue that the mythological creatures in the *Kalevala* were popular explanations for natural phenomena like the sun, the thunder, the Northern Lights, and the like. In what ultimately proved his most important methodological contribution, Aspelin suggested that a close comparison of redactions might enable scholars to determine "the natural phenomenon on which each tale is based." Aspelin believed he could explain the genesis of the poems psychologically and perhaps find "evidence

Newspaper readers in the Helsinki Folk Library, 1880s. A. Federley's drawing.

for Darwin's excellent theory of evolution." No further volumes in Aspelin's projected series of *Kalevala* studies were forthcoming, however, possibly because Krohn was simultaneously moving into the same research area.

After Borenius-Lähteenkorva had published his theory of the poems' "wandering" from West to East, Krohn dedicated himself to the study of the poetic redactions, and published his results in the first volume of *Suomalaisen kirjallisuuden vaiheet* ("The Stages of Finnish Literature") on the *Kalevala* (1883-1885).[110] This was the projected three-volume work discussed above, only the first volume of which appeared in Krohn's lifetime, the second and third on the *Kanteletar* and modern Finnish literature being brought out posthumously by Krohn's son Kaarle. Krohn began the *Kalevala* volume with an aesthetic analysis based on the lectures he had given at the Helsinki Academy for Women in the mid-1870s;[111] but in later chapters Krohn broke away from Lönnrot's *Kalevala* and expanded his scope to cover folk poetry in general.

Krohn's lengthy research caused him to revise his conception of folk poetry substantially, with the result that he was not entirely satisfied with the work he had done; as he says in that first volume, "a new perusal of the entire collection of poems [is] essential." The magnitude of the task that was to face folklorists of the next few decades had begun to dawn on the first of them.

In the second part of his volume Krohn culls abundant illustrations from the original Finnish folk poems in order to compare them with epic poems from other cultures, and then comes to the central section of his study, where he asks: How and where did the Finnish folk poems originate? How did they evolve into their present form? How are their parallels with other cultures' poems to be explained? Given the assumptions of the day and the reigning idea of his own dissertation twenty years before, Krohn's postulation is bold: "The poetic materials of the *Kalevala* came to us largely from the neighboring cultures." Poems travel in space, and comparisons of various versions both within a single culture and among many cultures reveal geographical progressions. Poems also change over time, although in the case of folk poetry, due to the recent collection of the poems, this is harder to document. These two evolutionary dimensions yield the method's "geographical-historical" name: according to Krohn the systematic application of the method should make it possible to reconstruct a reliable picture of a poem's evolution and origin. Krohn argues that the poems primarily originated in myth, less frequently in history, and that they traveled primarily from south to north and from west to east. This latter conclusion is based on Krohn's discovery that poems collected in more southerly and westerly regions were simpler and plainer than those collected in more northerly and easterly regions.

Krohn's theory met with immediate resistance from his colleague Ahlqvist, who in *Kalevalan karjalaisuus* defended Lönnrot's claims that the folk poems on which the *Kalevala* was based had originated in Karelia, and had not merely been collected there. Ahlqvist's evidence was largely philological, based on morphological and syntactical features like affixes and inflections.

On another central point, however, Ahlqvist agreed with Krohn. It was still generally believed at this time that the Karelians had possessed a complete unified cycle of epic poems; Ahlqvist joined Krohn in insisting that the *Kalevala* had received what coherence it had "at Lönnrot's hand, not that of its Karelian singers."

Twentieth-century scholars are in total agreement on this latter point, which K. A. Gottlund and Krohn had stressed, and as I mentioned earlier, the most recent work on the *Kalevala* is beginning once again to examine the text as a specifically literary creation. On the former point, the eastern or western origins of the poem, there is still no agreement, as the many symposia held in 1985 to celebrate the 150th anniversary of the *Kalevala* showed.

Twentieth-century folklorists have also outgrown Krohn's geographical-historical method, although they recognize the validity of the scholarly insight on which it was based. Krohn was able to apply Darwinian ideas to the study of folklore while not losing sight of the specific nature of the cultural phenomenon he was studying. The notion of the two directions of the evolution of oral tradition, temporal and spatial, is brilliant in its simplicity and clarity. Krohn knew the value of his observation and accepted Ahlqvist's predictable attack with equanimity. This is evident in a letter he wrote to his mother in 1886 after learning that Ahlqvist intended to contest his research results and prove the Karelian origins of the *Kalevala*:

> It can't be done with a snap of the fingers; you need more than a pompous proclamation. That fact alone would make me a Doubting Thomas, that he doesn't know the redactions, without which no scholarly *Kalevala* scholarship is possible. In any case, I am glad he is writing his critique, for it will draw attention to the issue and will thus facilitate the discovery of the truth, whatever it may be. (Quoted in Trast 67)

Valfrid Vasenius and the Study of Nordic Literature

After Kaarlo Bergbom's dissertation defense in 1868, it will be recalled, there was a ten-year hiatus before the next literature dissertation was defended. The next doctoral candidates to write their dissertations under Estlander were Valfrid Vasenius and Eliel Aspelin, both born in the 1840s like Bergbom, though slightly younger than him. They defended their dissertations in the year before and the year after 1880, and their scholarly careers mark a clear transition from the nationalist idealism of the earlier generation to a more exact, scientifically based positivism. Their work is itself transitional in that it contains new positivistic elements, but is still steeped in the romantic ideas of the nationalists.

Valfrid Vasenius (1848-1928) began his career in the university library, and in fact kept his job there until the turn of the century.[112] In the 1870s he edited the national bibliography *Suomalainen kirjallisuus 1544-1877* ("Finnish Literature 1544-1877," 1877), which drew on the library collections to supplement Pipping's earlier work, and which Vasenius himself later supplemented to reach the end of the century. In addition to his work in the library, Vasenius worked actively to develop the operating procedures of public libraries, wrote school textbooks on the Swedish language and literary history, pronounced on pedagogical issues, and later, in the early twentieth century, when the new Parliament was being set up, was consulted as an expert on (and proponent of) a proportionate electoral system. In the 1870s and 1880s Vasenius was an ideological radical who, as a reviewer for the *Helsingfors Dagblad* introduced modern European literature to Finnish readers, and who helped found and edit the Finnish-language periodical *Valvoja* (The Overseer), a journal dedicated to the introduction of modern social and scientific views into Finland. Vasenius was familiar with John Stuart Mill's and Herbert Spencer's ideas while still at university, and Darwin's influence on his thinking is clear in all his work on literature.

Vasenius wrote his dissertation on Henrik Ibsen, a 178-page study entitled *Henrik Ibsens dramatiska diktning i dess första skede* ("The First Phase of Henrik Ibsen's Dramatic Writing," 1879). There he examined Ibsen's early plays in search of the ideological concerns that would inform his major and more controversial plays. Vasenius discusses the public debate over Ibsen's plays at some length, passionately refuting Ibsen's detractors' accusations that he is inconsistent and socially irresponsible. He defends his own ideological method on the grounds that there is not yet enough data for the study of a modern author's "internal history;" his ideological alignments, on the other hand, are clear to all of his contemporaries, and thus are easier to study. Vasenius reaches the fairly general conclusion that in his early historical plays Ibsen emphasized the ethical impulse, and that his later ideology is already evident: his protagonists have in one way or another become alienated from themselves, stopped listening to the voice of their inner selves, and thus have become inwardly conflicted.

Three years after the dissertation, in 1882, Vasenius published a monograph in Sweden covering Ibsen's entire career, entitled simply *Henrik Ibsen.* This book resembles the dissertation in its attack on Ibsen's misinformed critics: in opposition to Brandes's readings of Ibsen, for example, Vasenius asserts that Ibsen was not at all a "preachy" writer; among recent ideas only evolution was important to him (327, 329, 339, 341). In fact, Vasenius places evolutionary thought at the very core of Ibsen's work, particularly the ideological evolutionism that believed in the displacement of outdated ideologies by more advanced ones. Individualism is central to Ibsen's plays as well, Vasenius argues, in the sense that the individual's responsibility for his or her own actions is always crucial to Ibsen. Contrary to popular opinion, conventionalism and love are also essential features in Ibsen's work. All this, Vasenius argues, disproves the popular notion that Ibsen writes in search of ideals and reconciliation without finding them. That notion only reveals readers' unwillingness to let go of outdated ideas and embrace that "revolution of the human spirit" that Ibsen calls for. Ibsen's guiding principle is that "the evolutionary work of love is man's greatest happiness" and also the greatest guarantor of human development.

These studies of Ibsen reflect the fundamental conflict that will characterize all of Vasenius's scholarly work. On the one hand, he attempts to examine Ibsen's art through the use of modern scientific concepts, above all evolution; but on the other hand he charges these positivistic methods with an idealistic vision. As Wilhelm Bolin wrote in his 1883 *Finsk Tidskrift* review of the *Ibsen* monograph, Vasenius sets himself the task of analyzing Ibsen's plays by gathering "together all the fragments of the dramatic characters' personalities," and thus of providing Ibsen's reader with a far greater pleasure than was possible "through a lens that misleadingly refracted the rays emitted by the sun of Ibsen's genius."

Valfrid Vasenius (M)

In 1880 Vasenius was appointed docent in aesthetics and literary history, simultaneously with Eliel Aspelin's appointment in aesthetics and art history. Thus the department that had limped along with Estlander for its only instructor received a strong injection of new scholarly blood. During this first tenure as docent, which lasted until 1887, Vasenius lectured regularly on aesthetics, less regularly on topics like poetics and the principles of aesthetic criticism applied to the plays of Schiller (1881), the history of Finnish literature (1883), the nineteenth-century history of Finnish and Swedish literature (1883-1884), and the analysis of drama (1887). In his teaching he adopted the seminar mode, which his student, later professor of literature, Werner Söderhjelm, recalls as follows:

> The new docent adopted a totally new style. In the first place he was much younger [than Estlander], and he immediately became our friend, was interested in our studies, and expressed his ideas on literature and other matters to us openly. In addition he arranged his first lecture course in such a way that his students could participate in it: as I recall, we were reading some play of Schiller's, and each of us could comment in turn on what we'd read, following which the instructor would summarize, correct, and supplement. In that first course there were

> only four of us, myself and three others, Hjalmar Neiglick, Petrus Nordman, and Alex Lundström, who later became a rector and publisher; but we were extremely taken with this novel teaching style, and it opened up for us an entirely new way of understanding literature. ("Vasenius" 371)

Other byproducts of Vasenius's interest in the Nordic cultures were his popular textbooks, *Lärobok i Sveriges och Finlands litteraturhistoria* ("Textbook on The History of Swedish and Finnish Literature," 1886), *Historiska undersökningar rörande Sveriges äldsta originalroman* ("Historical Investigations into the Oldest Original Swedish Novel," 1892) – a study of a novel written by Jacob Mörk in the 1740s – and, in a sense, the "Historiska undersökningar om Jacob Frese" ("Historical Investigations into Jacob Frese," 1884) as well, a lengthy study of a poet who had moved from Finland to Sweden in the early eighteenth century. In all of these works Vasenius's interest in the ideological contents and contexts of the works is evident, as is his unwillingness to remove himself to the rarefied realm of abstract aesthetic analysis.

Indeed, in his book on Frese he explicitly inveigles against the "abstract aesthetic" interpretation of literary texts, and grounds Frese's poems solidly in the thought of the period. His method is patently modern. He claims that it is easier to understand Frese's poems biographically when one explores the ways in which they express both the poet's time and his own life. Based on detailed archival sources and on close ideological readings of the poems, Vasenius places Frese's work in the specific ideological context of Pietism, a movement that shaped not only Frese but the entire period.[113] A more personal point is Vasenius's insistence that Frese was no sycophant to Carl XII, as Atterbom and Cygnaeus had suggested, but a poetic democrat and mouthpiece of his people's sufferings.

In the book on the oldest Swedish novel, *Adalriks och Giöthildas äfventyr*, Vasenius attempted to determine the extent to which the novel written in 1743-1744 was grounded in contemporary opinion on Sweden's position and fate after the loss of the war with Russia (and considerable land), and to which, on the other hand, it was charged with the social ideals of the so-called "period of freedom." Vasenius chides Swedish critics for their excessive aestheticization of this novel, and draws attention instead to the ways in which Mörk's social consciousness had been shaped by the freethinking ideologies of the previous decades. This, Vasenius claims, is most evident when one compares Mörk's novel with its foreign models, Fénélon, for example. Such comparisons, Vasenius argues, also demonstrate the utility of historical analysis as a foundation for aesthetic analysis. Despite Mörk's explicitly moral and political intentions, the free thought of the period gives a particular shape to the novel's artistic content: fleshes out its images, charges its mood with warmth, gives its characterizations consistency. From all this Vasenius draws a theoretical conclusion that one might characterize in terms of the primacy of historical context:

> Thus there is no other conclusion to be reached than that artistic activity too, to a much greater extent than is generally believed, is dependent on world view. And the natural conclusion to draw from this in turn is that an aesthetic appraisal of a literary work must be grounded in a

> thorough familiarity with the historical conditions under which it was created. (179)

He makes similar demands in his 1887 review of Lucina Hagman's monograph on Fredrika Bremer. Vasenius argues that in constructing a critical image of a writer, one must collect the major ideological features of that writer's work and arrange them so that the reader can judge the consistency of the writer's world view. One must demonstrate that all of the writer's works proceed consistently out of the same ideological principle, for example by "culling representative passages from the writer's works," thus allowing the writer "to depict and defend his own world view" (182-183).

This conception of the work of art as shaped by the world view of its period, which in turn is shaped by social conditions, informs Vasenius's readings of Runeberg as well. Back in the 1870s Vasenius had read two papers on Runeberg and collected them in *Om [on] Johan Ludvig Runeberg* (1877), and while their scholarly offerings are fairly slim, they contain the germ of Vasenius's later Runeberg criticism: in the linguistic and social conflicts raging in Finland, Finns might discover their own unity in the spirit of Runeberg. Vasenius returned to this theme in the 1890s, when Runeberg criticism was booming. It is revealing, given his position on the language question, that his first new monograph on Runeberg had already been accepted for publication in the Swedish Literature Society series when the increased polarization of the question at the January 1896 meeting of the Swedish Party induced him to withdraw his cooperation and promised collaboration on the society's *Finlands svenska vitterhet* series (Steinby 59).

Johan Ludvig Runeberg som konstnär ("Runeberg as Artist," 1896), which was intended as the first volume of a multivolume study of the poet, focused on Runeberg's use of the hexameter. Vasenius believed that Swedish prosody studies had come at the question of meter backwards, attempting to lay down rules for proper poetic writing rather than attempting to understand why poets wrote as they did. Vasenius's goal is no more nor less than the creation of a new theory of Swedish meter (11).

In accordance with his methodological vision, Vasenius seeks beneath a poem's outward form, i.e., its meter, the ideological exponent that shapes that form. Applying statistical methods to the *Elk Hunters* and Runeberg's manuscript revisions, he determines the "normal types" of Runeberg's hexameter and their deviations, abstracts out of them a new metrical rule, and tests its applicability on *Hanna* and *Julqvällen* ("Christmas Eve"). Thus he attempts to "control historically" what had been "discovered empirically" (112). Vasenius overgeneralizes to some extent from his specific discoveries to poetic meter in general, and his claims have been partially discredited by later scholarship. He argues that Runeberg's two hexametric foot-types shape not only his linguistic material but also his intellectual and emotional content: the three-syllable foot is inevitably found in passages of little significance, whereas the two-syllable foot signals the presence of powerful feelings or great conflicts. Thus "linguistic and schematic formalism," Vasenius claims, has lost its significance in prosody studies; meter is no longer a mere extension of grammar, but part of a general theory of art.

When Estlander retired in 1900, Vasenius applied for the position and wrote a second study of Runeberg, *Världsbild och poesi*, as his professorial dissertation. As the title, "World View and Poetry," suggests, it begins in Vasenius's now-familiar insistence on the importance of ideological analyses, but it also reflects a new attempt to supplement positivistic principles with new and dialectically opposed demands of the spirit – with the unsurprising result that Vasenius's vision is steeped in prepositivistic idealism. The result is a contradictory mixture of the positivistic emphasis on biography and the romantic emphasis on genius; the theoretical section, for example, is insistently positivistic in thrust, while the reading of Runeberg's works tends powerfully toward an idealistic ethical vision that implicitly refutes and is refuted by the theory (see Huuhtanen 100-102).

Vasenius devotes nearly half of the 143-page dissertation to an extended definition of world view: what is the factor, he asks, that combines the fragments of reality an individual perceives into a mentally coherent world view? A factor of the sort Vasenius is looking for should take the new empirically discovered laws into consideration while yet remaining a sufficiently comprehensive descriptor (130). Vasenius finds such a factor in the work of the fashionable contemporary philosopher Wilhelm Ostwald: the concepts of "energy" and "energeticity," which Ostwald claims allow us to integrate the formerly opposed realms of matter and spirit in a single scientific category. Vasenius proclaims that "the concept of energy as biologists have developed it thus meets all the scientific requirements set for the most comprehensive concept." The principle is positivistic in origin, but when applied to literature it leads to untenable speculation.

Vasenius's leading idea is a strong individualism, a myth or a cult of the genius, which he attempts to lash firmly to historical context through a kind of social-commission theory: the individual genius of a great writer is fundamentally the comprehension and expression of national individuality. The critic too must attempt to identify with this genius:

> The aesthetician must learn to feel what the genius felt, respect what he believed, desire what he desired. This is no more impossible than the poet's identification with the spiritual life of his people and time, in the midst of which he has lived, to which he has belonged. (130)

In drawing connections between the national idealism of Runeberg's time and Runeberg's texts Vasenius makes several valuable observations. For example, he notes that while Runeberg's *Kungarne på Salamis* ("The Kings of Salamis") is a classical treatment of a classical theme, the "inward life" of the poem is modern and topical: it is the relation between power and justice, which was a crucial issue for both Runeberg and the Finnish people in his period (130).

Similar notions reappear in the work published in Finnish, *Runeberg Suomen kansan runoilijana* ("Runeberg as the Poet of the Finnish People," 1903), but now the romantic-idealistic-nationalistic tones are stronger still. Vasenius stresses the unity between national mentality and individual artistic creativity as follows:

> If the great poet gives his people and all humanity a great deal, surely he has also received a great deal from them as well. And if we correctly apprehend this interplay of giving and receiving, we will perceive that

> the poet is the representative of his people and all mankind; in him we learn to know the highest ideological aspirations and needs of our people and of all people. It is then that we can most clearly understand our own position in regard to those aspirations. (1-2)

When Vasenius had studied foreign literary texts, his work had largely been received positively, with at most cautious criticism;[114] his Runeberg books, caught as they were between the old and the new schools, drew considerably more fire. This was only partly because of the internal contradictions of his work. A large part of it also was the fact that Runeberg was just then in the process of being canonized as the national poet of Finland, and any new scientific theories of literature that undermined or otherwise changed the image of Runeberg that was taking shape were felt to be inappropriate.[115]

While the vacant post left by Estlander was being filled, Vasenius had resumed his docent lectures, dealing with German and Swedish literature; and when, to his great bitterness, he was passed over in the appointment, he continued on as docent until he was appointed professor of Finnish and Nordic literary history in 1902. He held this post until his retirement in 1908, lecturing most frequently on the history of Finnish and Swedish literature, with particular emphasis on influences and international comparisons.

Vasenius's last great research topic was Finland's other major Swedish-language writer of the nineteenth century, Zachris Topelius, who had personally entrusted Vasenius with both the editing of his collected works and the writing of his biography. Vasenius edited the collected works, running to 34 volumes, between 1899 and 1907, and the biography, which ran to six volumes, began to appear in 1912, with the last volume appearing posthumously in 1930. The Topelius biography, which has sometimes been called Vasenius's magnum opus, is an attempt to integrate all of the details of Topelius's life into his character and world view, and to ground his literary and other activities solidly in the internal conditions of Finland. In quest of the former objective Vasenius details Topelius's life, with only an occasional use of the Procrustean bed. In any case, Vasenius succeeded in fleshing out an entirely new image of his subject, not only as an author but as a journalist and professor of history, and he also made a significant contribution to the contemporary image of mid-century political and social development in Finland.

Vasenius was the greatest Finnish student of Nordic literature in the nineteenth century. The study of Nordic literature had been given a promising start in Finland in mid-century by the dissertations of Toppelius and Lagus, but then had dwindled down in the 1860s and 1870s to a trickle. Around 1880 it showed signs of new life. In addition to Vasenius's work on Ibsen, a series of articles published in the *Finsk Tidskrift* in particular gave Nordic literature a new visibility. The articles dealt with such writers as Lidner (1876), Hans Christian Andersen (1877), Paludan-Müller (1878), Bjørnson (1880), Ibsen (1880, 1882), Strindberg (1880, 1883), Snoilsky (1881, 1882), Kielland (1883), and af Wirsén (1886). Many of these were written by Estlander, but Wilhelm Bolin was quite active in this field as well, though his major interests were German literature and Shakespeare. Quite a bit was written on Fredrika Bremer as well, including the monograph by Lucina Hagman, with its explicitly feminist content.

Werner Söderhjelm's 1884 dissertation on Schlegel falls chronologically into this same Nordic literature boom in Finland, but Söderhjelm was younger than the above-mentioned critics, and his work must be left for a later chapter.

The 1901 Appointment

The search for a replacement for Estlander, which began in 1898 and was finally concluded in early 1901, proved extremely difficult. The main problem was the interpretation of the post's field of specialization. Of the three applicants, Vasenius was above all a literary critic who used aesthetic theory in his literary criticism; Tikkanen was a pure art historian; and only Aspelin combined literary criticism with aesthetics and art history.

The definitional difficulties had two prongs, one for each field specified in the job title. Should aesthetics be considered a distinct branch of scholarship, for example, which would mean that a qualified candidate must be an aesthetician; or was it enough that a candidate should have applied aesthetics to the study of literature or art?[116] And what *was* "modern literature," anyway? The three referees appointed by the division of history and linguistics,[117] Henrik Schück, Estlander, and Werner Söderhjelm, interpreted "modern literature" in a variety of ways. The Swedish referee Schück implicitly included Finnish literature written in both Finnish and Swedish and the whole of Nordic literature in the field of "modern literature." Estlander was more restrictive: he argued that literatures already represented by existing academic chairs should not be included in the "modern literature" of this chair. This would mean excluding Oriental literatures, Russian literature, and the literatures covered by the chairs of Finnish and Swedish. Estlander was particularly adamant about this interpretation in his review of Aspelin's scholarly work, of which he would allow only the study of La Motte and criticism of German literature as "wholly within the purview of the chair in question." Since there was also a chair in Romance and Germanic literatures, held by none other than the third referee, Werner Söderhjelm, this was not an especially sensible or even self-consistent restriction. Söderhjelm argued from history: when the professorship was created, "modern literature" had meant the modern European literatures as opposed to the literature of classical antiquity. The position of Nordic literature had, Söderhjelm pointed out, remained somewhat unclear. Given that of all the forms of art only literature was explicitly mentioned in the title, and that literature had been the primary pedagogical focus during the entire history of the post, special emphasis should be placed on candidates' literary-critical qualifications. These varying turn-of-the-century definitions of the field covered by the chair are significant, as literary criticism and art history were just beginning to split into two fields in this period. Aesthetics, like critical theory in our own period, remained more amorphous in its disciplinary affinities. Literary criticism too was in motion, and would soon be making new demands on aesthetics.

The peer review of the three candidates' qualifications was contradictory. Estlander ranked Vasenius first, arguing that while Aspelin's work was more solid and methodologically stronger than Vasenius's, Vasenius was more profound and original a scholar and had a richer store of ideas than Aspelin, and Vasenius could be expected to develop the discipline in more fruitful ways. Estlander found Tikkanen's scholarly range too narrow and too peripheral to the chair, and ranked him third.

Schück was at a disadvantage in comparison with the Finnish referees, since he could only familiarize himself with scholarship written in Finnish by means of translated summaries. Most of Aspelin's work in particular remained alien to him for this reason. Schück admitted that, if the job title was interpreted literally, Vasenius would be the only qualified applicant; but he argued for a looser interpretation in line with the established practice of the chair, including aesthetics, literary history, and art history. Since each of the three candidates had published in at least one of these three areas, they must be ranked on the basis of scholarly merit. With this in mind Schück ranked Tikkanen first, Vasenius second, and Aspelin, whom he attacked for theoretical weakness, third.

Söderhjelm agreed with Schück in allowing publication in any one of the three areas – literary history, art history, or aesthetics – to qualify a candidate for the position, but insisted that, since literature had been central to the discipline from the beginning, literary-critical publications should be given special weight. On this basis Söderhjelm ranked Aspelin first: in addition to showing competence in all three of the chair's fields, he was the only candidate to possess a command of the chair's "most important area," the major literatures of modern Europe. One of the crucial factors in the appointment thus became *what* literatures one had studied: the French and German literature Aspelin had worked on was more prestigious than the Finnish and other Nordic literature Vasenius had worked on. As Söderhjelm phrased it, Aspelin was a scholar of *world* literature; Vasenius was more provincially concerned with the literature of Northern Europe. Söderhjelm also criticized Vasenius's "overrestrictive view of the tasks of literary-historical science," but still ranked him above Tikkanen.

Interpreting these rankings proved troublesome for both the division of history and linguistics and the university senate, which made the final recommendation. The division ranked Aspelin first, Vasenius second, and Tikkanen third (minutes, November 16, 1900); the senate then revised this ranking radically, placing Tikkanen first, Aspelin second, and Vasenius third (minutes, January 19, 1901). The senate debate indicates that the language question may have played a role in this decision. Professor of Oriental literatures Tallqvist argued that Aspelin's recent work on Finnish literature lacked a larger comparative framework. He granted that gathering and publishing material on a Finnish writer might have some national significance, but argued that it did not yet constitute literary history "in the modern scientific sense." This was not the place, Tallqvist argued, to reward patriotic endeavors; only scholarly credentials carried any weight. Certainly Estlander's low opinion of Aspelin had much to do with the latter's Fennophilia as well. The final appointment, however, rejected the senate's revision and returned to the division ranking: Aspelin was appointed professor of aesthetics and modern languages in April, 1901.

Eliel Aspelin

Much more clearly than Vasenius, Eliel Aspelin (1847-1917) placed himself in that disciplinary tradition that emphasized publication in literary criticism, aesthetics, and art history. He was the first significant scholar of Finnish art history,[118] and along with Julius Krohn he was a pioneer in Finnish-language literary criticism.

Aspelin was born the son of an associate pastor in Veteli, in Ostrobothnia.[119] After his graduation from high school in 1865 he moved to Helsinki to study aesthetics and modern literature, general history, Latin, Finnish, and philosophy, and took his Master's degree in 1871. His lifelong dedication to civic activities in cultural organizations began in his student years: by the end of the 1860s he was active in the group headed by Kaarlo Bergbom whose work would lead to the formation of the Finnish Theater.

He also took an early stance on the language question that was to have long-range effects. Considering it his duty to advance Finnish-language culture in every way possible, he chose literary texts written in Finnish as his first research topics, and gave his first paper at the annual meeting of the Ostrobothnian Students' Club in the fall on 1872 on Aleksis Kivi, the founder of Finnish-language literature. Kivi was then 38 years old, hopelessly insane, and in the last weeks of his life. Aspelin's paper, which was published later that same year in the *Kirjallinen Kuukauslehti*, was the first public account of Kivi's life, and it has become a cornerstone of biographical research on Kivi. A certain high idealism colors Aspelin's perceptive understanding of Kivi's realism, and heightens his sense of Kivi's significance as the spokesman of the Finnish-speaking populace in their own language, one who understands their positive core. In his emphasis on Kivi's Finnishness Aspelin expresses his disappointment that this great Finnish writer has had so few readers among the Finnish intelligentsia. This neglect is all the more unfortunate, Aspelin says, in that Kivi could have guided his readers to a true understanding of the Finnish character. The characters of Kivi's 1864 play *Heath Cobblers*, for example, had been "seized with a mighty hand from the ranks of the peasantry" (233), and its protagonist Esko resembled Kullervo from the *Kalevala*: both were "products of the same popular spirit" (234). The "Finnish eccentricity" of the seven brothers in Kivi's novel of that title, furthermore, was so great that only a Finnish reader could fully understand them. Kivi's one-act play *Lea*, whose plot this "son of the people" had adapted from the Bible, showed how deep the Finnish people's fear of God was (237).

In this emphasis on Kivi's role as the interpreter of the Finnish-speaking people's thoughts and feelings, and on Kivi's artistic significance as somehow the result or effect of the human significance of the people he portrayed, Aspelin laid the groundwork for a critical practice that later, during his tenure as professor in the early years of the new century, was to prove immensely influential. These were the boom years of Finnish-language culture and criticism, and Aspelin was to play an important role in their growth.

Having completed his university studies, Aspelin seems to have contemplated a teaching career: in the early 1870s he observed instruction in various subjects at the Normal Lyceum, and occasionally did hourly teaching there. But he was also strongly interested in art history, and was an active member

Eliel Aspelin-Haapkylä (M)

of the Finnish Antiquarian Society, which had been founded in 1870. He made two long research trips to Central and Southern Europe in the late 1870s, on the first of which, made on a university travel grant, he attended Herman Hettner's lectures in Dresden in February, 1876. Based on these trips he wrote and defended his Finnish-language dissertation on medieval triptychs in the fall of 1878. He later published other studies of art history, with a central emphasis on Finnish art; in fact, his scholarship was so broad in this field that the referees for the aesthetics and modern literature job considered him more qualified for an art history chair than a literature chair. In 1880 Aspelin was appointed docent in aesthetics and art history; and while Estlander lectured only in Swedish, Aspelin began to give his lectures in Finnish.

In the early 1880s, however, Aspelin had also begun to work on literary subjects, and when a second docent in aesthetics and art history was appointed in 1884 – J. J. Tikkanen – Aspelin began to turn increasingly to literature. His earliest literary criticism mainly dealt with foreign literature: after the book on the *Kalevala* mentioned in an earlier chapter (1882), he wrote "Lamottes afhandlingar om tragedin, granskade och jämförda med Lessing" ("La Motte's Treatments of Tragedy, Examined in Comparison with Lessing," 1886) and *Kansa Saksan kertomarunoudessa 1750-1850* ("The Folk in German Narrative Literature 1750-1850," 1894).

The 64-page study of Antoine de la Motte displays the general methodological striving for empirical exactitude that was typical of the 1880s. Aspelin defines his goals clearly, develops his argument carefully (though with rather scanty documentation), and formulates his conclusions precisely, with emphasis throughout on historical development. His primary sources are B.

Jullien's edition of the collected works of La Motte, published in Paris in 1859, including four theoretical articles on drama that Aspelin reads in terms of La Motte's position in "the history of aesthetics." Jullien's appraisal of La Motte's significance does not satisfy Aspelin; Jullien was still, Aspelin argues, the prisoner of the classical tradition. Jullien does not adequately support his case that the German reaction to French pseudoclassicism originated in France; he fails to make comparisons with German literature; and he fails to analyze the coherent impression left by La Motte's "paradoxes." Aspelin insists that La Motte was a brilliant critic of, indeed the first debunker of, pseudoclassicism, but no real innovator. He lacked sufficient vision of the new art's function for a true smashing of tradition. Aspelin's La Motte comes close to a new vision in his Voltairean emphasis on the importance of naturalness, but he is content to define naturalness as whatever pleased one's audience. Still, La Motte anticipates technical principles of the drama that would later become established, especially concerning dramatic exposition, dialogue, and internal coherence. Difficult as it is, due to the writers' different aims, to compare La Motte with Lessing and his *Hamburgische Dramaturgie*, Aspelin argues that Lessing follows roughly in La Motte's footsteps, complicating and enriching the French playwright's work.

In a preface to his book on the folk in German narrative poetry from 1750 to 1850, Aspelin says that one of his aims is "to shed an indirect light on the conditions for the birth and growth of our own country's literature" – and indeed throughout he grounds literature solidly in sociohistorical conditions. Aspelin focuses on the century during which the literary depiction of the "folk" or peasantry was just being developed, arguing that these depictions arose in the mid-eighteenth century out of the intensification of national and democratic thought, and were shifted into a new phase in the mid-nineteenth century by a marked improvement in the basic civil rights of the German peasantry – an improvement in which the literature Aspelin is concerned with played a significant role. In comparison with Aspelin's later, meticulously detailed studies of Finnish literature, this thin 156-page volume is marked by its clarity of vision, its lack of sentimentality, and its provocative argumentation – all of which would reappear in Aspelin's work only in the 1901 inaugural address.

Aspelin had been nominated for a professorship extraordinaire in aesthetics and modern literature in 1888; nothing came of the post then, but he was appointed to that chair in 1892. Having thitherto divided his teaching time fairly evenly between literature – including French literature (1883, 1889-1890), the *Kalevala* (1884-1885), German literature (1885-1886), and literary theory (1888, 1889) – and art history, upon his appointment he began to concentrate his teaching on literature and aesthetics. During his tenure as adjunct professor he lectured on the history of German literature (1892-1893, 1894, 1895-1896, 1898, 1899), the history of English literature (1894, 1895-1896), and the nineteenth-century history of French literature (1897), and gave introductory courses on literary history and aesthetics.

In addition to Aspelin, Vasenius, and Tikkanen, two younger scholars had applied for Estlander's job upon his retirement in 1898: Yrjö Hirn and J. O. E. Donner, both of whom had withdrawn their applications shortly after submitting them. This was the first time the new appointment procedure was put

Students between lectures in the university corridors. A. Federley's drawing.

into effect, eliminating the professorial dissertation, which previously had always been written specifically for each job application, as the criterial test of each candidate's qualifications. From this point on, candidates' qualifications were evaluated in terms of their scholarly work in its entirety.[120] Due to uncertainty as to whether the new procedure would affect this job search, Vasenius submitted a professorial dissertation anyway, his *Världsbild och poesi*, which was so riddled with internal contradictions that it only muddied the already murky waters.

Aspelin had written his biography of Lars Stenbäck, the Finnish Pietist minister and poet, in time for it to be considered in the professorial search: it appeared in Swedish in 1900 and in Finnish in 1901. This 600-page book was the longest literary-critical work Aspelin ever wrote; he explained its length on the grounds that he desired to do justice to Stenbäck not only as a poet but as a Pietist leader – as the clergyman who ultimately silenced the poet in himself. Aspelin's approach to Stenbäck – striving for objectivity, avoiding evaluation and judgment – is perhaps best summed up in the preface to the Swedish edition, where he declares his desire to portray both the poet and the Pietist "without interposing inhibitive criticisms" (3). Aspelin's programmatic Fennophilia is evident in the preface to the Finnish edition, according to which the Swedish-speaking Stenbäck was "a pure Finn in heart and soul," despite his estrangement from the Finnish-speaking people, which Aspelin ascribes to "our peculiar historical circumstances." In his presentation of the biographical data Aspelin tends to flood the reader with detailed factual information in chronological order without interpreting, explaining, or drawing connections with Stenbäck's literary work, with the result that the book has little value as literary criticism. Still, precisely in its wealth of raw biographical data the book offers a valuable portrait of early- and mid-nineteenth-century academic and Pietistic life in Finland.

As a scholar Aspelin was neither much of a theoretician nor particularly interested in detailed philological research. His clearest theoretical statement is made in his 1901 inaugural address, "On the Function of Literary History." There he at once argues for a dialectical approach to literature, which would see literary texts at once as shaping and shaped by society, and reaffirms his earlier romantic belief that literature is the expression of the *Volksgeist*:

> Because, in the clichéd but still pointed phrase, poetry is the flower of civilization, we require first that the critic determine those political, social, and cultural-historical conditions in which the poet lived and wrote, and show how his roots were in the people and what his position was in historical evolution; second, we wish to understand the poet's personality and philosophy of life, with all its sources and influences, in order the better to fathom what unique and independent role he played in, what contribution he made to, the national development; and third, we seek a detailed knowledge of the poet's work, the artistic form his genius and special character gave to what had been concealed in the *Volksgeist* or lain dormant in the realm of sheer possibility.

Aspelin reviews the development of literary historiography in France and Germany (Bernays, Gervinus, La Harpe, Hettner, the Schlegels, Taine) and warns against the philological direction in which some German scholars, like

Scherer, have taken *Germanistik*, and then comes to his own critical program. In his formulation the literary historian must be three things at once: a broad-based historian able to explore the appearance of the *Volksgeist* and the "idea" in a wide variety of phenomena; a psychologist able to fathom "the poet's temperament"; and an aesthetician with a grounding in empirical psychology capable of making aesthetic evaluations. In addition, lest he divorce art from life, the literary historian must be able to make ethical judgments.

There are allusions to new turn-of-the-century theoretical trends in Aspelin's address, notably his defense of subjective experience and its empirical psychological analysis; but in the main the piece harks back to a broad range of nineteenth-century literary criticism. But then, Aspelin had never brought theoretical concerns to the fore in his criticism. He was a close inductive reader of literature who, unlike Estlander, never had any difficulty projecting himself sympathetically into the work of realists and naturalists like Aleksis Kivi, Juhani Aho, Emile Zola, and Henrik Ibsen.

Aspelin, who changed his name to Aspelin-Haapkylä during the period in which it became fashionable to Finnicize Swedish names, retired from the professorship in 1908. The major scholarly project of his tenure as professor was a four-volume history of the Finnish Theater, which appeared between 1906 and 1910. In a total of 1600 pages Aspelin provides a wealth of documentary material to flesh out the development of Finland's national stage over the thirty-year period from its founding to the time of writing. The work also serves as a kind of biography of Kaarlo and Emilia Bergbom, who had directed the theater for its entire history.

His professorial lectures continued along the same tracks as before, with the central lectures on the history of German (1903-1904, 1904-1905, 1905-1906, 1907-1908) and French literature (1901-1902, 1902-1903, 1905-1906, 1906-1907) occasionally supplemented by more theoretical courses, such as one on the history of aesthetics and criticism (1902-1903). He was not a particularly inspiring teacher; as his student V. A. Koskenniemi, himself a professor of literature and a poet, later described him, "his elocution was . . . somewhat monotonous . . . with every word carefully inscribed in his small, rococo script on paper, as he stepped in his diplomatic jacket and sharply creased trousers to the podium."[121] Aspelin seems to have recognized his inability to reach his young students. After the last lecture he gave as a professor he wrote in his diary (December 1, 1907) that he had made no farewell speech, since "students pay so little attention to lectures that no close contact ever forms between them and their instructor" (143).

During his retirement Aspelin-Haapkylä published a three-volume collection of biographical remembrances of late-nineteenth-century Finnish writers, artists, and other cultural figures (1911-1914). He wrote almost entirely from memory, only occasionally drawing on secondary sources. One of the interesting minibiographies in the series was of the early Finnish prose writer and publisher Kaarle Jaakko Gummerus (3:151-303).

Aspelin played as active a role in advancing Finnish-language cultural institutions as Estlander had in the Swedish-speaking camp. He was one of the cofounders in 1890 of the Finnish-language publishing house Otava, served from 1902 to 1917 as director of the Finnish Literature Society, and, feeling that the Finnish Society of Sciences and Letters (Finska Vetenskaps-Societeten) was too heavily dominated by Swedish speakers, he helped found the Finnish Academy of Science (Suomalainen Tiedeakatemia), and served as the Academy's first chairperson from 1908 to 1917.

Aspelin did not hold the professorship long enough to have much of an effect on the theoretical and methodological development of literary criticism. During his tenure it was largely a group of younger scholars, especially the Swedish-speaking Werner Söderhjelm, Yrjö Hirn, and Gunnar Castrén, who were instrumental in transforming Finnish literary criticism for the new century – primarily ridding it of the last traces of nineteenth-century national idealism.

Literary Historiography in the Late Nineteenth Century

One of the significant features of literary studies in the last decades of the nineteenth century, especially the 1880s, was a widespread interest in historical surveys of literature. Krohn's history of Finnish literature, which remained unfinished at his death, was a prime example of this interest; but it was only one example. Finnish literary historiography advanced on many other fronts as well. In addition to the increased concern with evolution and its conditions, which was taking firm root in scientific thinking across the board, there was also an increased need for school textbooks on the history of literature. The first major histories of both Finnish and foreign literatures were written in the 1880s – though the first comprehensive survey of world literature did not appear until the new century, and even then only partially at the hand of Finnish scholars.

With certain reservations we might call historian Ernst Palmén's (1849-1919) 138-page history of the Finnish Literature Society, on the occasion of the society's fiftieth anniversay (1881), the first history of modern Finnish-language literature. Obviously, since Palmén is concerned with the operations of the society, he deals with Finnish literature only indirectly, insofar as it was published by the society. Palmén places particular emphasis on how the literary use of Finnish and the role of Finnish in society had changed – what he calls in his title "the progress of Fennicism." In this context he does devote some rather idealized Fennophile pages to original and translated Finnish literature, and to original Finnish nonfictional writing.

A similar work in certain respects was the 82-page essay-collection *Studier i finska vitterheten efter 1830* ("Studies in Finnish Literature After 1830," 1883) by the Helsinki teacher Albin Lönnbeck (1856-1914). Like Palmén, Lönnbeck discusses literature written in Finnish over a fifty-year period, with a very brief survey of the period preceding 1830. Unlike Palmén, however, Lönnbeck approaches literature written in Finnish somewhat from the outside, with the rather critical if nonetheless perceptive eye of a dedicated Svetophile.[122] Lönnbeck was primarily an art historian, defending his dissertation in 1888 on ancient Greek relief art; but he was also one of the first Finnish scholars of Wecksell, and in the 1890s he edited the third edition of Wecksell's collected works, appending a lengthy biography of the poet to the series.

The 95-page *Oppikirja suomalaisen kirjallisuuden historiassa* published the next year (1884, and in English translation under the title *Handbook of the History of Finnish Literature* in 1896) by Bernhard Fredrik Godenhjelm (1840-1912) was the first history of Finnish literature written specifically for school use. Godenhjelm had long been a lecturer in German at the university and was a vocal advocate of Finnish women's education, and he conceived both the literary history and his other textbook, on poetics (1891), in strictly practical pedagogical terms. The heavily idealistic thrust of his approach to literature is most clearly evident in his magnum opus, the 514-page *Runous ja runouden muodot* ("Poetry and Its Forms"), which was written in the 1880s and 1890s but was published posthumously in 1914. It was the most extensive Finnish poetics prior to Rafael Koskimies's *Yleinen runousoppi* ("General Poetics," 1934).

Godenhjelm's aesthetic mentor was the romantic idealist Vischer, and the first section of his posthumous poetics (21-192, which had appeared as a separate piece in 1885, under the title "Runous kuvausvoiman ja kansanhengen tuotteena" ["Poetry as a Product of the Imagination and the *Volksgeist*"]) is unmistakably Vischerian. This section also contains the first Finnish history of world literature: Godenhjelm covers Oriental and ancient Greek and Roman literature, medieval literature, Renaissance and Reformation literature, and modern Spanish, French, and German literature. In his discussion of each national literature Godenhjelm explores its national character and its interaction with other literatures, arguing that "true poetry is always universal," but that it embodies its universal ideals not through vague abstractions or generalizations, but rather in the form "given it by the imagination of each individual nation" (178-179). In his reflections on the national characteristics of Finnish literature, Godenhjelm launches a tradition that was to become dominant in the early twentieth century, depicting Finnish literature in terms of key phrases like "a fresh feel for nature," "a profound sentimentality," and "a natural simplicity of expression" (187). It is possible to draw a direct line from these phrases to the literary histories written by Viljo Tarkiainen and Rafael Koskimies.

The *Handbook of the History of Finnish Literature* still seethes with the nationalism of Snellman: "the history of Finnish literature is the history of Fennicism." Godenhjelm deals in passing with scholarship on the Finnish language and a few Swedish-speaking Finnish writers, and concludes with a brief look at the Finnish literature of the 1870s and 1880s, though he says that "the most recent period is not yet history and thus cannot be introduced in the same manner as the earlier periods."

Godenhjelm condensed his idealistic conception of literature into the 1891 textbook *Lyhyt runous-oppi. Jatko-opistoja varten* ("A Brief Poetics: For Teachers Training Colleges") as well. In art, Godenhjelm declares here, humans render visible "the world as it truly is, the world in the full rightness and elevation of its purpose." This ultimate truth is not phenomenal factuality; various ephemeral phenomena and even glaring defects may be true, but they are not "in accordance with *truth*, i.e., with the purpose of all being." Even natural beauty for Godenhjelm is incomplete and haphazard: "perfect beauty can only be produced by art," by compiling the fragmentary reflections of beauty and extracting from them a coherent, essential world. These notions were already anachronistic when Godenhjelm wrote them, but because they were used in the schools they had a good deal of influence, and several reprint editions of his books appeared around the turn of the century.

Valfrid Vasenius's Swedish-language textbook on the history of Swedish and Finnish literature appeared shortly after Godenhjelm's, in 1886. It was an attempt to portray the development of both Swedish and Finnish literature in parallel, from the Middle Ages to the nineteenth century. One sign that in the 1880s the writing of literary history was more conscious of methodology than before is Vasenius's defense, in his preface, of the historical as opposed to the formal approach to literature:

> I believe the teaching of literature should be historical rather than aesthetic. Abstract aesthetic critiques of writers and literary works must make way for accounts of literary evolution that offer as comprehensive an explanation as possible of the various trends of poetry, their origins and development, as natural and integral parts of their times. . . . Criticism must also show how poets express the national awareness of their times. Only in this way will we pass from the desert of aesthetic abstractions to the fresh fountain of noble patriotic thoughts and feelings, and will poets become the objects of students' love and admiration rather than of their cynical analysis. (4)

The two parts of this quotation reflect the theoretical doublethink we found characteristic of Vasenius's literary criticism: empirical historicism combined with national idealism. Vasenius goes on to define the concepts "literature," "national literature," and "literary history," insisting throughout that a historical approach to literature should always focus on "national literature," which is to say literature that most clearly reflects a people's deepest thoughts and feelings. Especially in the works of great writers, these are represented as the people's "idea," or as "an image of the objective toward which the people strive, and of the mode of life which the people consider optimum." Vasenius argues that the national literature that should be the historical focus of literary studies may include not only *belles lettres*, but also historiography, orations, social essays, and certain religious writings. Since national literature reflects the essence of the people at various points in their history, Vasenius continues, it is grounded in national history, and "literary history is best categorized in same way as national history."

As early as 1879 the Finnish Literature Society had begun to plan the publication of a comprehensive history of world literature;[123] but despite the best of intentions the society was unsuccessful at commissioning the writing, or even the translating, of such a history in the nineteenth century. A number of literary histories in various language regions were in fact begun. In 1861 Ludvig Leonard Laurén (1824-1884), a teacher from Vaasa, had written a rather modest school textbook in Swedish on the history of French literature, whose literary-historical value Snellman had placed under serious question in a review. A bad Swedish translation of Julian Schmidt's rather cursory survey of nineteenth-century English literature from the 1850s appeared in Turku in 1871. Better than both of these was the 111-page textbook Karl Viktor Bremer (1836-1884), a Helsinki teacher and journalist, wrote for the girls' school on the history of Swedish literature in 1874. Bremer draws most heavily on the Swedish literary histories written by Atterbom and Malmström, the textbooks

B. F. Godenhjelm (SKS)

written by Bjursten and Claëson, and Lagus's history of Finnish literature: Bremer deals not only with the literature of Sweden, but with Finnish literature written in Swedish as well. The 1880s saw the publication of two histories of Russian literature. One was written in German by a Hämeenlinna high school teacher, Edvard Wilhelm Palander (1845-1914), *Uebersicht der neueren russischen Literatur* (1880), and published as a 91-page addendum to the annual report of the Hämeenlinna Normal Lyceum. Palander had been born in Ingria, had gone to school in St. Petersburg and Viipuri, and had attended university in Tarto and Helsinki, where after taking his Master's degree in 1870 he had held a lectureship in Russian for a few years.[124] The other was written in Swedish by Georg Fraser, a 114-page study entitled *Vitterheten under nittonde seklets reformtid i Ryssland* ("Literature in the Nineteenth-Century Reform Period in Russia," 1888).

In general the literary histories of the 1880s continued to perpetuate, in their selection of texts, their periodizations, and their explicit argumentations, the mid-century notion of the organic unity of literature and the *Volksgeist*. In a scholarly sense, of course, this idea was already outdated, but these literary histories were written with another social function in mind. They were not interested in being theoretically up-do-date; their purpose was the more pragmatic one of serving the expanding nationalist movement in the schools, by so construing literature in historical perspective as to provide the new generations with a firmer national identity.

Other Late-Nineteenth-Century Critics

Significant literary criticism was written in the 1880s and 1890s outside the discipline of aesthetics and modern literature as well. Most philosophers, for example, thanks largely to the Hegelian and generally German idealist tradition, were deeply interested in literature and the other arts, and published critical studies of them. The professor extraordinaire of philosophy Wilhelm Bolin (1835-1924), for example, a well-known Feuerbach scholar of Swedish-German extraction who worked in the university library from the 1870s to 1912, was particularly interested in literature and the theater. He published numerous articles on such writers as Shakespeare and Ludwig Anzengruber in Finland and Germany; indeed, he was on the payroll at *Die Gegenwart* magazine for two decades, from 1876-1897. J. J. F. Perander (1838-1885), who was a docent and later professor of philosophy, later still professor of pedagogy and didactics, wrote extensively in the 1870s and 1880s on the theater and Finnish folk poetry. Thiodolf Rein (1838-1919) too, professor of philosophy and author of a massive biography of Snellman (1895, 1899), wrote numerous articles dealing with literary-theoretical and aesthetic issues.

Though Hjalmar Neiglick (1860-1889), who died at the young age of 29, only gave a few papers and published a few articles on literature, his contribution to the discipline was significant. He was the first in Finland to take an impassioned stance against the nineteenth-century idealist philosophers; indeed in 1916 Werner Söderhjelm called him the only "Finnish theoretician of the 1880s." At the annual meeting of the Uusimaa Students' Club in the spring of 1885 he gave a talk on "Voltaire och nittonde seklet" ("Voltaire and the Nineteenth Century"), drawing heavily on Henry Thomas Buckle's *History of Civilization in England.*[125] Here he sided strongly with the rationalist, anti-Christian Voltaire against the idealism of Schelling and the "insane dialectics" of Hegel, in so doing, of course, implicitly attacking also such influential Finnish Hegelians as Snellman. In conjunction with his attack on philosophical idealism Neiglick launches an attack on literary romanticism and offers a passionate defense of realism. As he put it, romanticism was an elitist and egotistical movement that the suffering European poor were in the process of smashing (29).

The philologists also frequently touched on literary-critical issues in their teaching and scholarship. The job title for the philologically conceived chair in the Finnish language mentioned literature as well, and in the last decades of the nineteenth century lecture courses were still regularly offered by this chairholder on Finnish literature. The art and literature of classical antiquity were of central importance to the department of classical philology, and the traditionally philological study of ancient Greek and Roman literature continued to be practiced well into the new century.[126] When Werner Söderhjelm was docent in Romance linguistics he lectured on the *Song of Roland* from the pragmatic perspective of language learning (1890-1891); during his tenure as professor extraordinaire he lectured on Molière and his times (1895-1896);

and while professor of Germanic and Romance philology he lectured on the history of medieval French literature (1897-1899), Dante and medieval Italian literature (1899-1900), and on many other literary topics in the new century. Johannes Wilhelm Öhquist (1861-1949), lecturer in German, frequently taught courses in German literature, and wrote numerous articles on literary topics for the *Finsk Tidskrift* and other journals. Professor of Oriental literature Strandman lectured on the older Arabic literature (1895-1896) and other philological topics.

Of particular significance was the teaching and research done in the field of Russian language and literature, where until the late 1890s the university had the only full professorship in a non-Finnish modern language. Even before the professorship was created there had been a dissertation defense in the field: Gabriel Geitlin (1804-1872) defended his 54-page dissertation on Lomonosov, *Dissertatio academica de meritis litterariis Lomonossovii*, in 1829 and 1830. Geitlin begins with a brief survey of Russian history and cultural history and biography of Lomonosov, then discusses his work, with particular emphasis on his Russian grammar book. Russian-born Jakov Grot (1812-1893), who had been appointed professor of Russian history, statistics, language, and literature at the University of Helsinki in 1841, worked all through the middle of the century to strengthen Finland's cultural ties to Russia.[127] He lectured on Russian literature, edited a Russian reader (*Rysk läsebok*, 1848), and wrote the first Finnish textbook on Russian history (1850). In addition he wrote regularly for the St. Petersburg-based journal *Sovremennik*, often introducing new works of Finnish literature, such as Runeberg's *Julqvällen* (1840) and *Nadeschda* (1841), or reporting on Finnish research trips to Russia, such as Lönnrot's journeys in northern Russia and the travel book he published about them (1843). Grot also edited the collected works of G. R. Derzhavin and, when he published them in St. Petersburg upon his return in 1853, he introduced them with a biography of Derzhavin.

The Finnish-born professor of Russian Arvid Frithiof Nordqvist (1828-1889) lectured regularly on Russian literature in the 1870s and 1880s, as did his successor, the Russian-born Viktor Semenoff (1858-1919), in the 1890s. Semenoff only held the professorship until 1896; in 1910 he became Assistant Director of the Ecclesiastical Committee in the Finnish Senate. In 1863 Nordqvist began a series in the *Litteraturblad* entitled "Teckningar från den slaviska litteraturen" ("Notes on Slavic Literature"), but the series broke off after the first installment on Adam Mickiewicz.

The greatest literary critic among the professors of Russian language and literature was Josef Mandelstam (1846-1911).[128] Mandelstam was born in Pultava, and when he arrived in Finland in 1897 to take up the professorship he could not speak a word of either Finnish or Swedish. He lectured in German at first, but later learned enough Swedish for teaching purposes. His extempore lectures were extremely popular, and one of his students, Eino Kalima, who later became a theater director, recalls that Mandelstam was one of the best teachers at the university (Aalto *Modern* 144). His articles on Russian literature appeared both in the Finnish-language journal *Valvoja* (his inaugural address on the major trends in nineteenth-century Russian literature, 1898) and in the Swedish-language journal *Ateneum* (on Tolstoy's literary works and their relation to his moral philosophy, 1898; on Russian writers' reflections on freedom of speech, 1899; and on Maksim Gorky, 1902). Sometimes an article he published in one journal would later be translated for publication in the other. Mandelstam lectured on the new Russian literature (Dostoevsky, Tolstoy, etc.) and Russian folk poetry (1898-1899), the history of Russian book reviewing (1900-1901), European romanticism and its influence on Russian poetry (1902-1903), the language of Pushkin and Lermontov (1903), Turgenev (1903-1904), Tolstoy (1904), late-nineteenth-century Russian poetry (1904-1905), the relationship between the literary language and the vernacular (1905-1906), and the history of Russian drama (1908). He also lectured frequently on the history of Russian literature from the eighteenth century to the present. His literary-critical magnum opus is the 405-page *O kharaktere gogolevskogo stilja* (1902), a stylistic analysis of Gogol's language in relation to the Russian vernacular on the one hand and the language of Gogol's contemporaries on the other. Mandelstam also does some psychological readings of Gogol's language in relation to his personality.

Literary criticism was also written outside the university, notably by school teachers. Both of the major literature societies, the Finnish Literature Society and the Swedish Literature Society in Finland, played an active role in encouraging the practical study of literature.

Wilhelm Lagus (1821-1909) began in mid-century as a student of ancient Greek and Roman literature and was later appointed professor first of Oriental languages, then (in 1866) of Greek literature, and finally became rector of the university. During his tenure as rector he devoted himself to the study of Finnish culture and literary history. His most important literary-critical work was the 360-page study of the Finnish period of the Swedish poet Johan Henrik Kellgren, *Skalden Johan Henrik Kellgrens finska lefnadsminnen* (1884). Lagus offers a detailed description of Kellgren's stay at the University of Turku from the 1770s to the 1780s and his work in Sweden thereafter, providing in the process a complex picture of literary life in Turku and Stockholm in the late eighteenth century. Lagus was a member of the first board of directors for the Swedish Literature Society in Finland, and was given the honor of publishing his edition of H. G. Porthan's letters to Mathias Calonius as the first volume of the society's new publication series (1886). Significantly, in his review of Lagus's book on Kellgren in the *Finsk Tidskrift* in 1885, Estlander called it a broad survey of "the dawning of our literary culture," meaning specifically our *Swedish*-speaking culture. This implicit association is made clear in the last sentence of Estlander's review, where he notes that when reading scholarship like this one feels the urgency of creating a literature society that could gather evidence of "svensk" or Swedish(-language) culture in Finland. Indeed, in a way these two works by Lagus had the effect of opening up the scholarly field that in the decades around the turn of the century began to focus on Finnish literature in Swedish and the earliest literary contacts between Finland and Sweden. The Swedish Literature Society in Finland was to play a central role in commissioning and publishing that scholarship.

Another industrious scholar who wrote in Swedish was Wilhelm's son (and Gabriel Lagus's nephew) Ernst Lagus (1859-1923).[129] Lagus taught school in Helsinki all his adult life, and wrote his doctoral dissertation on the history of classical language teaching in Finland (*Studier i den klassiska språkunder-*

visningens historie i Finland, 1890); but at the same time published numerous literary-critical articles in periodicals and in publication series brought out by the Swedish Literature Society, edited scholarly works and wrote two scholarly monographs. Of these various scholarly activities by far the most important were the editing jobs. Around the turn of the century he played an active role in the publication of Swedish-language Finnish folklore, brought out an edition of Porthan's letters in two volumes (1898, 1912), and edited the works of Mikael Choraeus, a Finnish poet who had moved to Sweden. In the first volume of his edition of Choraeus (1901), which appeared as volume two of the Swedish Literature Society's series *Finlands svenska vitterhet* ("Finnish Literature in Swedish"), he published the poems and a 190-page biography; in the second volume (1903) he grouped the letters, sermons, and prose. Choraeus had received some critical attention before Lagus, particularly from Finnish scholars. Franzén had published some of Choraeus's poems and a brief biography of him in Sweden in 1815; Gabriel Lagus had published a series of articles on him in 1868 in the *Finlands Allmänna Tidning*; and in his own 1875 monograph on Choraeus J. O. I. Rancken had uncovered new documentary material and a few lost poems. In addition, in the 1890s the Swedish poet Snoilsky had written a stirring biography of Choraeus just as the Academy of Sweden had decided to strike a medal commemorating him. But new sources had recently surfaced, including some poems, the sermons, and the particularly significant papers contained in the so-called Porthan-Tengström trunk, which was opened in 1900 in accordance with the specifications of Porthan's will. Drawing on these new materials Lagus was able to write a substantially more comprehensive biography than his predecessors, with a heavier emphasis on the life than the work.

Lagus's best-known monograph was an explication of Runeberg's *Songs of Ensign Stal* written for the general public (*Förklaringar till Fänrik Ståls sägner*, 1894). Here Lagus offered information on the poet's life, the 1808-1809 War of Finland, specialized terms used in the poems, and people and places mentioned in the poems. In the Runeberg boom of the turn of the century four editions were brought out of Lagus's booklet, and it was frequently used in schools.

Indeed, after his many scholarly editions Lagus focused exclusively on Runeberg, writing articles on him until the First World War. Many of these dealt with manuscript discoveries and were more or less indirectly tied to Lagus's hobby horse, the idea of founding a Runeberg museum. Perhaps the most significant was the 54-page article "Runebergs tidigare diktning, uppfattad af samtiden" ("Runeberg's Early Poetry, as Perceived by His Contemporaries," 1895), in which he explored the early reception of Runeberg's poems through a close look at magazines and newspapers.

After the History of World Literature project fell through, the Finnish Literature Society began to direct their publishing activities increasingly into the area of folkloristics, which overlapped in significant ways with literary studies as well. Otto Donner (1835-1909) had earlier published "Kalevipoeg jumalaistarulliselta ja historialliselta kannalta" ("Kalevipoeg as Myth and History," 1866) with the Society, and in the 1890s it was followed by Oskar Relander's (1836-1930) stylistic study *Kuvakielestä vanhemmassa suomalaisessa lyyrillisessä kansanrunoudessa* ("On Imagery in the Older Finnish

Lyrical Folk Poetry," 1894) and other folklore studies. The Society also published a series of biographies of nineteenth-century Finnish writers, A. H. Bergholm's book on Abraham Poppius (1899), Lauri Suomalainen's book on J. F. Lagervall (1901), and Emil Nervander's book on Fredrik Cygnaeus (1907), which had originally been published in Swedish (in 1892). The idea was to publish a series of literary histories and biographies roughly parallel to the rather more extensive one being published by the Swedish Literature Society, of which more later.

The biographical tradition in Finnish scholarship took firm hold in the early twentieth century, but it was grounded in work done over a long period throughout the nineteenth century. The central role played by cultural-historical biography in Finnish society is already evident in the great unfinished collection of 42 Finnish biographies from mid-century, *Finlands minnesvärde män* ("Finland's Notable Men," 1853-1857). The editors of the collection explain the cultural-historical criteria they have used in selecting the biographies to be written by saying that cultural history is far better suited to Finnish scholarship than political history:

> The political history of Finland is tied to Sweden and takes all its substance from thence; but the evolution and dissemination of culture has always been the accomplishment of each individual people, who have lived in a country and learned to love it; and the annals of history have amply shown that the Finns adopted and independently adapted that civilization that was brought to them from beyond their borders. (ii-iv)

Prior to the 1880s the most significant brief biographies were an anonymous one of Johan Frosterus (1820), Hipping's of Jacob Tengström (1843), Kellgren's (1851) and Vasenius's (1872) of Fabian Collan, Cygnaeus's of Robert Tengström (1853), Borg's (1853) and an anonymous one (1855) of M. A. Castrén, and Cygnaeus's of F. M. Franzén (1872). All of these ran from thirty to forty pages in length and appeared in periodicals either upon the subject's death or on a later anniversary of his death. Central book-length biographies would include Snellman's *Matias Alexander Castréns lefnadsteckning* ("Castrén's Biography," 1870), Fredrik Cygnaeus's *Johan Jacob Nervander* (1846), C. W. Törnegren's *Johan Lillienstedt* (1853-1854), and the popular biographies written in Finnish by Thiodolf Rein (*G. A. Wallinin elämä ja matkustukset* ["The Life and Travels of G. A. Wallin"], 1872) and Erik Edlund (*M. A. Castrénin elämä ja matkustukset* ["The Life and Travels of M. A. Castrén"], 1878).

Beginning in the mid-nineteenth century the Finnish biography began to evolve from short essayistic pieces in the popular press into a longer and more scholarly format, which begins to take hold toward the end of the century. The first significant successes in this genre were Eliel Aspelin's biographies of Johannes Takanen (1888) and Werner Holmberg (1890, both with the telling subtitle "Life and Works"), Thiodolf Rein's *Johan Vilhelm Snellman* (1895, 1899), Aspelin's *Lars Stenbäck* (1900), and Werner Söderhjelm's *Johan Ludvig Runeberg* (1904, 1906). These works already reflect the two directions into which Finnish biographies were beginning to branch off: in the historical-political biography political and economic activities were typically depicted in a moral and ethical light, while in literary- and cultural-historical biographies a kind of aesthetic opposition begins to raise its head, a rejection of bourgeois political and economic goals (cf. Scheuer 55).

The first real pioneer in the genre of the book-length Finnish biography, however, preceded the foregoing slightly: Johan Elias Strömborg's (1835-1900) "biographical notes" on Runeberg, which ran well over a thousand pages in length and was partly based on the older eulogistic tradition. Strömborg was a high school teacher from Porvoo who delivered a eulogy on Runeberg in 1878, in the first birthday commemoration of the poet after his death, and in what soon became a tradition continued delivering biographical orations on Runeberg nearly every year until his own death in 1900. He collected his orations and began publishing them in 1880, and the last volume appeared posthumously in 1901. Together they cover Runeberg's entire life and literary output, and constitute the first comprehensive biography of a Finnish author. Fortuitously, Strömborg's work began to appear in print just as a massive Runeberg cult was taking shape.[130]

Strömborg sought to provide the most extensive and detailed information on Runeberg's life and works then available. His list of newspaper articles dealing with Runeberg in both Finnish and foreign papers, for example, is impressive, as is his list of translations of the works. In addition to recording the publication of Runeberg criticism and translations, he writes his own critiques of Runeberg's works, occasionally even attacking previous readings. One such case is his interpretation of *Ensign Stal*, where he takes a stand in opposition to Cygnaeus. Strömborg is most thorough and also most vivacious in his depiction of Runeberg's first and last years, his childhood and decline. For the latter he had unique qualifications, since he lived in the same house as Runeberg and visited him almost daily, and his notes on these last years are so detailed that they were not published in their entirety until Karin Allardt brought out an expanded edition of the biography in 1927-1931.

PART FOUR
The Era of Quantitative Growth and Internationalization

Toward a Transitional Phase

The evolution of Finnish literary criticism at the turn of the century was guided both by methodological changes outside Finland and by institutional changes inside the university, specifically in the proportionate size of the various academic disciplines.

About the time Julius and Kaarle Krohn were developing the geographical-historical method in the late nineteenth century, folkloristics had split off into an independent discipline. In the teaching program for 1889-1890 Kaarle Krohn was listed for the first time (although he continued to lecture as before on literature as well) as a docent in Finnish and comparative folklore; and his appointment in 1898 and 1908 as, respectively, professor extraordinaire and full professor of Finnish and comparative folklore constituted major milestones in the discipline's move toward independence.[131] In terms of influence on the discipline of literary criticism, perhaps the most significant Finnish folklorist was Antti Aarne (1867-1925), who had studied in Russia in the 1890s; his concern with the motif structure of fairy tales offered important input to the literary study of motif as well. Another early folklorist whose work proved fruitful for Finnish literary criticism was A. R. Niemi (1869-1931), whose 1898 study of the composition of Lönnrot's *Runokokous Väinämöisestä* ("A Collection of Poems About Väinämöinen," a kind of trial run for the *Kalevala,*) was the first detailed analysis of Lönnrot's method for editing individual folk poems into a single autonomous poetic work.

Kaarle Krohn's teaching in the department of Finnish language and literature in the 1880s and early 1890s gave that discipline a heavy literary content as well: he lectured on Finnish metrics (1889), the history of modern Estonian literature (1889), the language of sixteenth-century writers (1890), and the history of Finnish literature overall (1890-1891, 1894), or beginning with Lönnrot (1891-1892, 1895-1896) or Agricola (1892-1893), or in its most recent phases (1894). When Arvid Genetz (1848-1915) succeeded August Ahlqvist as professor of Finnish, the significance in the discipline of research and teaching in literary criticism diminished markedly. Like his successor, E. N. Setälä, Genetz was a mainstream linguist – although, like his predecessor, he too was a poet. Also like Ahlqvist, he was a rigid poet-professor who utterly despised modern literary trends. Where Ahlqvist had condemned Aleksis Kivi's realism, Genetz (as a poet, under the pen name Arvi Jännes) condemned Minna Canth's naturalism. Even Kaarle Krohn's lectures on literature tended to focus, as the century drew to a close, on the *Kalevala*. It is highly significant for the history of Finnish literary criticism that linguistics (in its broad, pre-Saussurean sense) took over the entire discipline of Finnish in the late nineteenth century, for this generated a need for a new professorship in Finnish literature – a post that was not to become a reality until the new century.

The creation of a professorship in the modern languages in the 1890s also had an important impact on Finnish literary criticism. The need for research and teaching in the modern languages had long been recognized. Back in the 1860s August Ahlqvist had proposed the creation of a professorship in Germanic and Romance philology, and in 1877 Estlander had proposed the splitting of his own discipline. He argued that the broad disciplinary scope of the field required so many specialized methodologies – historical, linguistic, philosophical, critical – that it was becoming increasingly impossible for one person to maintain competence in all of them. Adapting the French model, he proposed the creation of two separate chairs: one in modern literature (nyare litteratur), which would include modern literary history and Germanic and Romance philology, and another in aesthetics and art history, which would cover the theory and history of literature and other forms of art.[132]

Neither proposal was acted upon. The transition to a more specialized disciplinary structure began in 1894, when Werner Söderhjelm was appointed professor extraordinaire of Romance philology; in 1898 he was to be appointed to the newly created full professorship in Germanic and Romance philology. In his 1894 inaugural address, which was published under the title *Moderna språk och deras studium i Finland* ("Modern Languages and Their Study in Finland," 1894), Söderhjelm divided the discipline into two branches, the linguistic and the literary, calling the latter the study of "language's finest manifestation" (6). Since the division of history and linguistics required modern language majors to minor in aesthetics and modern literature, in practice the study of modern languages and modern literature went hand in hand. Literary criticism increasingly shifted toward German and the Romance languages, philologists became increasingly interested in literature, and Söderhjelm, himself by training and inclination a literary scholar, began to have an impact on dissertation topics. To be sure, here in the early 1890s Söderhjelm still insisted on the special nature of literary studies within a philological discipline:

> Here too [in literary studies] the philologist must apply the methods learned in his language studies; hence it is best to lead him, via the detailed and careful study of the work of some writer, not so much into the evaluation of its aesthetic content or the analysis of its formal and thematic structure, as into its contextualization in contemporary culture, the determination of its historical significance and relation to other contemporary literatures. (28-29)

This was also Söderhjelm's own methodological program, one that was apparent in his own scholarly work and that was to have a significant impact on Finnish literary criticism, as he shaped a generation of literary critics beginning in the 1890s, ultimately as the first appointee to the new chair in Finnish and comparative literary history in 1913.

Since the "prescientific" debates over a national literature in the 1840s, both theoretical research into literature and methodological discussion had lain relatively fallow, for various reasons: the number of active critics in Finland was small, and they had little turnover. C. G. Estlander, after all, held the only full professorship in literature for over three decades. This situation began to change at the turn of the century. A new interest in critical methodology sprang up. The idealistic conception of literature, theoretically grounded in Hegel's philosophy and Vischer's aesthetics but based emotionally in a chauvinistic desire to display the national Finnish nature of the literature written in the country, had dominated literary criticism until the 1880s. It was still influential, and even received new impetus from resistance to Russification; but at the same time new radical ideas began to manifest themselves in the Swedish-speaking elite beginning in the 1870s. Swedish-speaking intellectuals began to turn to foreign models like Brandes, Ibsen, Strindberg, and Zola for positivistic ammunition against the national idealists. This movement was most evident at an early stage in the scholarly work of Valfrid Vasenius and Werner Söderhjelm; but it is important to note that, just as the last representatives of the idealist tradition were already incorporating positivistic elements into their work, so too were the new positivists still heavily steeped in the idealistic tradition.

The general trend of this intense new phase in Finnish literary criticism was a scientism based on the natural sciences, along with an increase in theoretical and methodological reflection. If B. F. Godenhjelm's 1891 poetics was an anachronistic rehash of ideas that had been current in the sixties and seventies, the new positivism began to raise its head in the late seventies. In his 1879 review of Robert Prölss's aesthetics text, for example, Wilhelm Bolin expressed notions that were to be symptomatic of the late-nineteenth-century debate over the science of literature. Bolin praised Prölss for his "concrete" approach to aesthetics, the absence in his text of "metaphysical baggage" and "mysticism," and predicted that aesthetics was entering a transitional stage that would lead to a solider epistemological base (165). In her 1886 *Valvoja* introduction to Taine's environmental theory, Hanna Andersin revealed some reservations about its more programmatic features but in general characterized it as an important theory central to the times. Andersin praised Taine for his freedom from national and religious prejudice and his simplicity and naturalness, "all of which puts his work in diametric opposition to the theories of the German aestheticians, who attempt to subordinate artistic phenomena to metaphysical laws."

These trends in Finland ran parallel to those in Sweden, where a similar move took place from the earlier aesthetic speculation on literature to a new grounding of literature in cultural history. The central Swedish representative of this shift, Henrik Schück, argued that speculative nineteenth-century aesthetics had given literary studies an ahistorical, even anachronistic conception of earlier writers by viewing them through the "lens of philosophy."[133] Finnish academics followed Swedish developments closely, especially through the *Finsk Tidskrift*: in his 1886 review of Schück's book on Shakespeare, for example, Wilhelm Bolin praised Schück's critical and rigorously historical approach, and lauded him for throwing over that "vaporous [svamlande] worship" that had once been considered the duty of a cultured person, to the point where one had begun to doubt one's own powers of immediate comprehension.

The turn of the century saw the introduction of numerous new scientific principles for the study of literature, driven by sociological, psychological, or philological concerns and typically grounded in causal thinking. In his 1897 defense of his sociopsychological theory of the origins of art in *Valvoja*, Yrjö Hirn admitted the great significance of Darwinism as the source of a new way of thinking that stressed the "unity of all mankind"; Gunnar Castrén opened his 1902 dissertation defense with a polemic for a new exact psychology of art;[134] and in his 1909 article on Runeberg's *Kung Fjalar* ("King Fjalar") Valfrid Vasenius first posited and then proved a hypothesis in order to demonstrate that even in interpretations of poetry hypotheses can be proven "as bindingly as any other scientific hypothesis in any other scientific discipline" (270-271).

The halcyon days of literary science at the turn of the century were bordered on both sides by significant Finnish translations: in 1887, at the urging of Minna Canth, a booklet translating excerpts from Georg Brandes's literary history *Hovedstrømninger i det nittende Aarhundredes Literatur* was brought out; and in 1915 the *Philosophie de l'art* written by Brandes's mentor Hippolyte Taine a half-century earlier appeared in Finnish translation.

The striking increase in the number of Finnish literary works also had the effect of opening up new scholarly possibilities. In a sense Finnish literature had "broken into" the academic world in 1894, when Estlander chose Juhani Aho's controversial naturalist novel *Papin rouva* ("A Minister's Wife," 1893) as his sample text in the piece he wrote for the invitation to an M.A. graduation, entitled "Om förhållandet mellan konsten och moralen" ("On the Relation Between Art and Morality," 1894). Pondering the writer's ethical and moral responsibilities, Estlander chastised Aho (and, by implication, all of the new literature) for failing to take a clear moral stance on the eponymous minister's wife whose extramarital affair formed the center of his novel's action.

Despite the quantitative growth of Finnish literature, its relative youth was often cited in the 1890s as an explanation for its relative invisibility in literature courses at the university. In his 1895 defense of the role played by studies in aesthetics and literature in the university, Eliel Aspelin wrote:

> Now as regards . . . the discussion of Finnish literature in university lectures, it has been and in part still is impossible. Our literature is so young that even its founders have scarcely all departed this life. Its treatment is difficult for other reasons as well. Unless the instructor has plenty of free time to seek out the profundities of our literature in sundry scattered places, it will not be lectured on until many scholars working in concert have uncovered its foundations.

Still, by the turn of the century such a substantial body of Finnish literature existed that it offered a promising field for scholarship. In addition, various political pressures – the Russian government's attempts to "unify" or annex Finland, the 1899-1905 "years of oppression" during which Finnish resistance to unification was branded "separatism" and "mutiny" – provided increased motivation for the forging of a strong national identity. Alongside the new anti-idealistic positivistic approaches to literature, then, the earlier social imperative toward national idealism continued to inform much literary criticism. Indeed one of the central tasks for literary criticism seemed to be the discovery of literary idols for popular emulation, and all of the older critics (from Estlander to Vasenius and Aspelin) and many of the younger ones as well did their part in this project. The central figure in this search for heroes was Runeberg, but before long a Finnish-speaking hero was needed as well, and one was found in Aleksis Kivi.

Methodological and theoretical reform of literary criticism at the turn of the century was facilitated by the formation of literary circles frequented by both writers and critics. These included the *Euterpe* circle, formed around the periodical of that name, the university library circle (of which more later), and the student fine arts society "Studenternas förening för konst och litteratur."[135] Between 1901 and 1903 this last included primarily university scholars and students and a few writers, and its discussions tended to revolve around the relation between art and morality. Opinion was sharply divided over the artist's moral responsibility and the relationship between art and reality, but the idealistic position seemed to be in the ascendancy over the Tainean environmental theory. In the heavily lecture-oriented university studies of the day, the arts society constituted what might be thought of as a supplementary seminar, where topical scientific issues and positions spread quickly, even beyond the confines of the university to creative art.

Werner Söderhjelm

Werner Söderhjelm (1859-1931) was part of the generation of intellectuals who grew up in the idealistic climate of Runeberg and Topelius, but who adopted the new radical ideas of the 1880s.[136] He was born in Viipuri the son of a lawyer, and took his Master's degree in 1882. His early career reflects the ideological watershed of the times in a number of ways. His school teachers included Fredrik Polén, a veteran of the Fennomania movement, and he studied at the university under the idealistic aesthetician Estlander – but also under the more modern Vasenius. Fundamentally Söderhjelm was fairly conservative and loyal to his roots, as is evident in his scholarly attachment to Runeberg: traveling abroad to pursue his studies in 1883 he said his goal was to learn modern research methods in order to study the work of Runeberg.[137]

In the 1880s Söderhjelm found influential scientific mentors in the Danish scholar Georg Brandes and the Swedish scholar Henrik Schück, both of whom were later to become his personal friends. Most of his postgraduate studies, however, with a heavy dose of philology, were undertaken in Central Europe – Vienna, Munich, Berlin, and Paris.

Beginning in the 1870s the dominant status of Vischer's aesthetics was beginning to fade in Central and Northern Europe, and to be displaced by a new philological and historical method largely developed by Wilhelm Scherer. Since theory-laden aesthetic and philosophical speculation was alien to Söderhjelm's temperament, he found the new approach attractive.

In the spring of 1883, then, Söderhjelm traveled to Munich to begin his postgraduate studies. There he worked with Franz Muncker, on whose advice he chose for his dissertation topic the early-eighteenth-century playwright Johann Elias Schlegel; Konrad Hoffmann, who taught him modern philological methodology; and above all Michael Bernays, a prestigious representative of the Scherer school who taught him the significance and the possibilities of detailed textual criticism. In his memoirs, *Läroår i främmande länder* ("Student Years in Foreign Lands," 1928), Söderhjelm characterized Scherer's and Bernays' method as follows:

> Like Bernays, Scherer believed that German literature was closely tied to other aspects of German culture; he defended the application of philological and historical methods to modern as well as classical literature; he condemned the use of philosophy to study literature, but paid close attention to the origins of a poetic work, even when they were clearly and purely based in personal experience. (55)

Söderhjelm defended his 138-page doctoral dissertation, *Om Johann Elias Schlegel, särskildt som lustspelsdiktare* ("Schlegel, especially as Writer of Comedies"), in Helsinki in the fall of 1884. This particular Schlegel provided an excellent subject for a scholarly project dedicated to the overthrow of national idealism in favor of a more international, comparative approach. At the age of 24 Schlegel had moved from Germany to Denmark, where he had played an active part in literary life; and cosmopolitan influences were to shape his work in significant ways all through his life, to the point where he had all but been written out of German and Danish national-idealist literary histories.

Söderhjelm ties Schlegel's early work to the ideas of his Leipzig teacher Gottsched, but also sees him as a precursor of later German classicism. Indeed one of the central methodological principles of the dissertation is the importance of literary genealogies:

> For one must assume that it is of the utmost importance to trace every step in the progression whose outcome is something of the order of German classicism – to follow the trail of the men among whom the new ideas first blossomed, and to determine the foreign influences that led their thinking in freer, more open directions. (i)

The dissertation focuses on Schlegel's greatest work, his comedies, but Söderhjelm deals in passing with Schlegel's full range of literary and other activities in an attempt to flesh out a comprehensive portrait of his "literary development, his independent thought, and the foreign trends that shaped his opinions" (ii). He argues for the primacy of a historicizing approach in this sort of study; and the key concepts of his argument are "model," "influence," and "development."

Söderhjelm sent Schück a copy of his dissertation for comment, and Schück wrote back to say that he was particularly pleased by the "truly scientific" method Söderhjelm had employed. He thought he detected, he remarked, the influence of German literary science, of "scientists who have finally learned to construct literary history on a foundation of philology and history rather than aesthetics."[138]

In the summer of 1885 Söderhjelm traveled to Vienna, where he briefly studied literature under the well-known Schererian Erich Schmidt and Romance philology under Adolfo Mussafia.[139] In the fall he traveled on to Paris, where his teacher was the famous scholar of language and literature, Gaston Paris. On Paris's advice Söderhjelm began to study French Renaissance literature, which was later to become one of his major research fields. One early article that grew out of his association with Paris was "Sur l'identité du Thomas auteur de Horn" (1886), in which Söderhjelm used close textual analysis to argue that, contrary to the opinions of scholars like Francisque Michel, the same-named authors of the Tristan and Horn novels were in fact different people. Paris also became an important methodological model for Söderhjelm, who soon insisted that Paris's blend of philology, aesthetics, and psychology was the only truly comprehensive way to understand a literary work.[140] In his memoirs, Söderhjelm recalls Paris's ideas with admiration:

> Like all historical facts, literary products are phenomena contingent upon special conditions. The learned man's task is to comprehend these phenomena in all their varying forms, determine the moment of manifestation and significance of each, explain their interrelatedness, and ultimately trace out of them general laws. (178)

And as a scholar Söderhjelm characterized him as follows:

> It seems to me that many features in Gaston Paris's work and personality – his frank and open spirit, his astonishing intellectual flexibility, his unflagging industry – might reveal him to have been a student of Renan's. On the other hand his absolute freedom from vague philosophical speculation and his literary-historical method demonstrate his close familiarity with the aspects recommended by Taine for this sort of research. (178)

The comparative approach held pride of place in the studies that immediately followed Söderhjelm's dissertation. In an 18-page article entitled "Zu den Plautinischen Einflüssen bei Holberg" ("On the Influence of Plautus in Holberg," 1885), Söderhjelm compares the Danish author's play *Didrik Menschenkraek* with Plautus's play *Pseudolus*, from which Holberg stole his plot almost unchanged. Söderhjelm argues that Holberg adapted Plautus's plot to reigning social conditions and even made certain dramaturgical improvements, but failed to flesh out certain characters as powerfully as Plautus. Söderhjelm also assumes a programmatic position on influence studies, arguing that the critic is to progress no further "than the author himself had done, and to discover influences only in writers the author in question had actually known" (2).

In his 44-page study "Petrarca in der deutschen Dichtung" (1886), Söderhjelm explores the apparently scanty connections between German literature and the early Italian Renaissance. Due to the German "master craftsman" mentality, Söderhjelm argues, the great literature of the Italian Renaissance did not find a foothold in Germany until the mid-eighteenth century, when a wave of translating, studying, and adapting Renaissance literature began. Whereas one can trace the German reception of Dante and Boccaccio in broad outline by listing translations, the situation is quite different with Petrarch,

Werner Söderhjelm (HKM)

who, Söderhjelm insists, had a much more vital influence on German literature. He sets himself the task of exploring the Petrarchan elements in the work of German poets who called themselves Petrarchans. After a review of the important German discoverers of Petrarch from Meinhard (1764) to Herder (1791-1795), and another of Petrarch translations and imitations from Gleim (1764) to Bürger (1789), Söderhjelm turns to a close formal analysis of the language of August Wilhelm Schlegel's translations of Petrarch. He shows how Schlegel shifted from early free translations toward strict prosodic, metric, and thematic equivalence, and how this shift affected the artistic success of the translations: in the earlier renditions Schlegel captured more of the "fresh poetic tone" of Petrarch's sonnets than he did in the later.

In February, 1886, Söderhjelm had been appointed docent in modern literature, but his lectures early that year only dealt with literature philologically. Söderhjelm was already planning work on Runeberg and medieval European literature, but due to the academic job structure of the 1880s he decided to take Estlander's advice and shift his research interests further into philology, where a professorship might be forthcoming in the near future.

Indeed, based on his experiences in Germany Söderhjelm himself contributed to the development of philology in the university, writing an article for *Valvoja* in 1886 deploring the low standard of modern language instruction both in the schools and in the university and calling for the creation of a new chair in philology.

After some further philological study in Berlin in the late 1880s Söderhjelm was appointed docent in Romance philology in 1889, professor extraordinaire of Romance philology in 1894, professor of Germanic and Romance philology in 1898, and, when the chair was divided in 1908, professor of Romance philology. Throughout his philological career, however, Söderhjelm kept up with literary science, and a significant proportion of both his teaching and his research dealt with literature. His lectures on literary subjects included: Molière and his time (1895-1897, 1903-1904, 1911-1912), the history of French literature in the transition from the Middle Ages to the modern period (1897-1899, 1902-1903, 1905-1906), Dante and medieval Italian literature (1899-1901), Goethe's *Faust* (1901-1902), and the history of realism in modern French literature (1912-1913). That he also kept up with literature in Finnish is evident in his five Swedish translations of Juhani Aho novels, including *Papin rouva* ("A Minister's Wife," 1893). And in early 1895 Söderhjelm had written his mother:

> I have now begun – but this must for the time being remain between you and me – serious research on Runeberg, which I hope will produce more useful and lasting results than all the linguistic articles and studies that really should be written by native speakers of the languages in question. The work I'm planning now has been on my mind for several years, but I have left it alone because I felt it was too peripheral to my true field; now however I think I can devote just enough of my free time to Runeberg, without neglecting my work, to write a book on him – or perhaps several, for the subject is vast.[141]

Söderhjelm first applied his philologicohistorical method to Finnish literature in his 1900 article "Anteckningar om och ur handskrifterna till Runebergs lyriska dikter" ("Notes on Runeberg's Lyrical Poems based on his Manuscripts") – the most thoroughgoing study of Finnish poetic manuscripts ever undertaken.

Söderhjelm's central work on literary subjects consists of three biographies and a period study. The first biography was *Karl August Tavaststjerna* (1900), a friend of Söderhjelm's youth and an important late-nineteenth-century Swedish-speaking Finnish prose writer. The second, often called Söderhjelm's magnum opus, was the two-volume *Johan Ludvig Runeberg* (1904, 1906), with the subtitle that Söderhjelm was the first literary biographer to use and that was to become almost a cliché of biographical writing, "hans lif och hans diktning," his life and his poetry. The third biography was of a contemporary Swedish writer, also in two volumes, *Oscar Levertin* (1914, 1917). The period study, *Åboromantiken* ("Turku Romanticism," 1915), was a historical analysis of Finnish literary life in Turku in the 1810s and 1820s.

All three biographies are structured after the same pattern: first a thorough depiction of the historical context, both material and intellectual; then the writer's life, works, and their reception. The works are tied closely to the life, with particular attention paid to comparative perspectives, especially demonstrable influences on theme and form. In the Tavaststjerna biography, for example, Söderhjelm says his aim is to determine the interrelations of the author's work and "his time, his environment, and his own personality" (1-2). This biography was written about a close friend only two years after his death, and it is considerably more personal than the other two; in the second edition (1913) Söderhjelm attempts to objectify his analyses by pruning off the more personal remarks. Both here and in the Levertin biography, on the other hand, Söderhjelm is careful to respect the reputations of those friends and relatives of his subject who are still living.

The 1000-page Runeberg biography too, a second edition of which appeared in 1929, was heavily influenced by the demands of the time in which it was written. It was commissioned by the Swedish Literature Society in Finland, and while Söderhjelm sought to create a totally new portrait of his subject, one unbound by tradition, the times would not allow such a radical deviation from convention. As Gunnar Castrén recalls, the biography was written "for a generation who looked up to Runeberg with almost blind worship, and it was written in a time during which resistance to the Russian years of oppression elevated his patriotic value, if that is possible, to an even higher level than before" (*Humanister* 40). Though Söderhjelm had, for example, done a detailed analysis of the biographical background for "Svartsjukans nätter," he makes only veiled reference to that background in the biography; and his insinuations that Runeberg's emotional life might not have been quite so harmonious as it had earlier been depicted – that the great poet might have harbored passionate feelings for other women – provoked instant hostility in ardent supporters of the Runeberg cult.[142]

Throughout the biography Söderhjelm insists on the importance of studying an author's cultural environment as a key to understanding his work. In Runeberg's case this means determining the environmental factors that shaped the birth and development of the poet's patriotism and literary ambitions and work, and how in turn the works influenced Finnish thought. Söderhjelm has certain reservations about treating Runeberg as a product of his environment, but continues:

> But this poetry still looms against a background without which it cannot be understood; it is connected with so many other parallel events and expressions of the time, forces that were shaping nineteenth-century Finland; it develops in an intellectual climate favorable to it. All this must be taken into consideration in a discussion of literary history, and only in this light can Runeberg's exceptional national significance be henceforward grasped. (1:2-3)

Söderhjelm insisted on seeing nineteenth-century Finnish literature in the context of European cultural trends, and called Runeberg's work "the mighty northernmost branch" of classicizing, romantic, and national trends especially in Germany. This emphasis manifests itself in the biography in the form of frequent analytical and evaluative comparisons with Central and Northern European literature. On the other hand Söderhjelm calls Runeberg an early independent "realist" who gives his poems birth on the solid foundation of reality, and who participates with all his heart in the life of the people. This last, Söderhjelm suggests, also helps explain the positive response Runeberg got from the Finnish people and the massive influence he had on the formation of a Finnish identity.

The Runeberg biography also has something of the air of a cultural reckoning. Söderhjelm predicts that Runeberg's native Swedish may lose its significance as "the true carrier of culture in Finland" (633), and his importance to Finns will thereafter lie more in his national character than in his language.

The publication of Söderhjelm's biography of Runeberg provoked a large-scale debate among Finnish and Swedish literary scholars, revolving centrally around the question of Runeberg's models. Söderhjelm himself was unwilling to identify as a model any work whose influence over Runeberg's writing could not be shown philologically. After he had completed the first volume of the biography in 1903, for example, he wrote an article in *Euterpe* criticizing, among others, Otto Sylwan's attribution of influence on Runeberg to Tegnér (327-328). The strongest proponent of Runeberg's foreign influences was a docent from Uppsala named Johan Mortensen, who devoted two long articles in 1904 and 1908 to the thesis that Runeberg's poetry was heavily influenced by German classicism (Herder, Voss, Goethe) and Walter Scott's "realism." He argued that Söderhjelm's philological method could only show certain individual linguistic congruences, no true intellectual indebtedness. In his 1907 review of both volumes of the biography, Otto Sylwan tentatively supported Mortensen's conclusions. In his own methodological polemics in 1905 and 1912, Söderhjelm warned against the danger of speculation in influence studies and insisted on remaining within the realm of empirical proof. He was, nonetheless, ready to admit that Runeberg's influences still offered a highly fruitful arena for further study.

The debate makes clear not only that Söderhjelm was more of an empiricist than his Swedish colleagues, but also that, like other Finnish Runeberg scholars, he was more inclined to perceive the great national poet as a unique, autonomous genius. This was not necessarily the simple result of the Romantic genius theory, as Swedish critics argued; it also attested to the continuing power of the nationalist interpretive tradition in Finland. According to Söderhjelm critics who harped on Runeberg's foreign models had not understood the ways in which "everyday conditions and trends in Finland" could inspire a writer, and what might be internalized by a "patriotic spirit and populist temperament" of Runeberg's caliber. Runeberg could be a patriot, Söderhjelm insisted, without having learned it from neo-Gothic romantics or Walter Scott.[143]

The study of the literary center in Turku in the early nineteenth century, *Åboromantiken*, focuses on ideological trends, as the subtitle indicates: "och dess samband med utländska idéströmningar" (and its connection with foreign ideological trends). Söderhjelm's book is clearly influenced heavily by the Swedish scholar Anton Blanck's work *Den nordiska renässansen i sjuttonhundratalets litteratur* ("The Nordic Renaissance in Eighteenth-Century Literature"), which he uses to identify literary and ideological models for early-nineteenth-century Finnish literature in preromantic work of the late eighteenth century. His aim is to show, first of all, the close ties between the intellectual movements that characterized academic life in Turku from the 1760s to the 1828 move to Helsinki and general intellectual trends in Europe, and second, how these ideas in their Finnish forms then became central to the development of Finnish culture. Turku was, Söderhjelm argues, the idea center for the North, the "cradle of civilization" (199). He explains the immense significance of romantic ideas in Finland by the fact that they arrived in the country at a time when everything was in the process of being remade from scratch, which left romanticism a clear field in which to operate.

Like Blanck, Söderhjelm views Porthan's work rather simplistically from or through a nationalist perspective: Porthan's *De Poësi Fennica* and a host of earlier phenomena are treated as the preparatory stage for and thus beginnings of Finnish "national romanticism." This prejudice survived well into the twentieth century – see Paul van Tieghem's *Le Préromantisme* (1948), for example – and it was not until Annamari Sarajas's 1956 dissertation that Finnish scholarship was awakened from its century-long "romantic sleep" (Sarajas 1-4).

Despite this overemphasis on romanticism in early-nineteenth-century literature – indeed, in nineteenth-century Finnish culture in general, since Söderhjelm argues that romanticism was relatively unaffected by the realistic reaction and continued strong into the new century – *Åboromantiken* nevertheless offers a clear picture of the Turku litteratis' ideological roots beginning in Herder and Porthan.

Söderhjelm wrote his last literary monograph or "essay" on an internationally known Finnish-language writer. *Johannes Linnankoski* (1918, trans. into Finnish 1919) was Söderhjelm's only book written on Finnish-language literature; it was intended for a Nordic readership, and can be seen in the context of the Russian revolution and the Finnish situation in the winter of 1918 after the Civil War. For Söderhjelm Linnankoski personifies the best qualities of the Finnish people; his active work in public education met all the standards that Söderhjelm had set himself as a young man in relation to the "people" (education, cultural work, Europeanization), and that after the bloody conclusion of the Civil War seemed more essential than ever.

Alongside his work on Finnish literature, Söderhjelm continued to work on Renaissance literature, in 1907 coauthoring a popular book on the Italian Renaissance with his brother Torsten (Werner wrote five of the book's nine chapters, Torsten four).[144] The Söderhjelms advanced a thesis that was soon discredited: that Renaissance man had once and for all discarded his barbaric past and grown into full humanism. Torsten Söderhjelm (1879-1907), who died while still in his twenties, was a competent scholar of French literature. He was a philologist by training, wrote his dissertation on Old French, but felt closer to art, literature, and the theater, all of which he wrote articles on in *Euterpe* and elsewhere. These articles were published posthumously in a collection edited by his brother Werner and Gunnar Castrén entitled *Uppsatser och kritiker* (1908).

For decades Söderhjelm had been planning to write on the fifteenth-century French short story writer Antoine de La Sale, a topic suggested to him by Gaston Paris. His original idea, as Gunnar Castrén reports in *Humanister och humaniora* ("Humanists and Humanities," 21), was to write a book-length biographical and literary-historical study; but this project never came to fruition. Only a few shorter pieces appeared: the 27-page "Spuren von Ciceros verlorenem Traktate De Virtutibus bei einem französischen Schriftsteller des fünfzehnten Jahrhunderts" ("Traces of Cicero's Lost Treatise *De Virtutibus* in the Work of a French Writer of the Fifteenth Century," 1904) and the longer, 152-page study "Notes sur Antoine de La Sale et ses oeuvres"

("Notes on Antoine de La Sale and His Work," 1908). Both studies show the importance in La Sale's work of allusions to Valerius Maximus and other Roman writers.

A few years later Söderhjelm returned to roughly the same ground in a book published in Paris, *La nouvelle francaise au XV:e siècle* ("The French Short Story in the Fifteenth Century," 1910), and in a popular Swedish version of the same work later that same year, *Studier i fransk berättarkonst* ("Studies in French Narrative Art"). In these studies of older French short story writing, based heavily on Söderhjelm's own earlier research, he paid particular attention to late-medieval French "realism." The Renaissance marked the beginning of the modern era, in which, unlike the Middle Ages, people thrilled to reality rather than fantasy, life deposed abstract theorizing, perception deposed speculation, the individual took precedence over public convention, and in literature and art a new "realistic age" began. Söderhjelm defends his choice of topics on the grounds that the short story clearly reveals the two faces of the French Renaissance: on the one hand the spontaneous national aspect, on the other the borrowed Italian aspect. Though Söderhjelm is primarily concerned with the artistic features of the short stories, he is also interested in the symptomatic value and evidential force of the works as social products. For him literature is "a reflection of the *Volksgeist*"; the fifteenth-century short story reflects emergent bourgeois society. Typical of Söderhjelm's method are the recurring key concepts "kinship," "model," "equivalence," "influence." He argues that the stories he examines demonstrate the great age of the modern French short story: a sharp perception of reality, a close and penetrating psychological analysis, dramatic force, a concentrated certainty of expression, a sense of humor, and many typical stylistic features, such as the use of antithesis, all tie the fifteenth-century short story to the present, even though the tradition is not unbroken over the intervening four hundred years.

The two 1910 books also reveal, in the nooks and crannies of Söderhjelm's explicit positivistic empiricism, hints of his implicit national idealism. In the Swedish edition, for example, he compares the lyric poetry that so suits the introspective German temperament to the short story that better suits the socially grounded French temperament, and argues that a purely aesthetic appraisal "contributes to the history of the literary manifestations and devices of the French spirit" (19).

Söderhjelm's argumentation also reveals his more practical motives, however, in particular his polemical stance in opposition to Yrjö Hirn's aesthetics. One must, he pontificates, decisively defend the critic's right and duty to examine the artwork itself, against those "more or less authoritative voices" who with bombast and demagogery would subordinate aesthetics to moral philosophy. These sideswipes at Hirn and defenses of the philologicohistorical method may be read in part against the backdrop of the upcoming appointment to the chair in aesthetics and modern literature; and indeed in 1910 Söderhjelm swallowed a bitter pill indeed in losing out to the "moral-philosophical" aesthetician Hirn.

The *Francesco Maria Molza* monograph that appeared the year after the appointment, in 1911, might be described as the Söderhjelm brothers' last collaboration. When his dissertation had been approved Torsten had traveled to Italy to research the *Italiensk renässans* book; and when that appeared he began to collect material for a biography of the late-medieval Italian writer Francesco Maria Molza. Before he had completed his research, however, he died in Florence, in January, 1907. Werner Söderhjelm finished and wrote up the research, including a wealth of detailed information on both Molza and the Italian Renaissance.

After his defeat at Hirn's hands in 1910, Söderhjelm still dreamed of a professorship in literary studies, and in 1913 he got one – one created especially for him, in "inhemsk och allmän litteraturhistoria" (domestic and general literary history). He served in this post until Finland gained its independence and he was named Finnish ambassador to Sweden. In his inaugural address in 1913, published the same year in *Valvoja*, he gathered together a number of his most central ideas about a field of scholarship dear to him, Finnish literature written in Swedish. Appealing to Tainean environmental principles, he argued that Swedish-language Finnish literature could not be assimilated to or identified with Swedish literature. It was as "Swedish" as American literature was "English":

> While it is linguistically one with the literature created by Swedes (though there too there are certain differences), it has not remained congruent ideologically, any more than people who spend their lives in different political circumstances, surrounded by different natural conditions, in different social environments, will retain all the same character traits and philosophy of life. (590)

He went on to reiterate his long-held position that now excellently motivated the field specification of the new professorship: the literature of each individual country must always be studied in an international literary context. The concept of "inhemsk" or "domestic" literature mentioned in the job title also fits Söderhjelm's sense of Finnish literary scholarship: the Finnish literature written in each "domestic" language must be studied in relation to the other.

During his tenure in the professorship Söderhjelm lectured on the history of world literature until 1800 and European literary trends since 1800, with particular emphasis on Scandinavian and Finnish literature (1913-1914), Runeberg's early work (1913-1914), European literary trends in the nineteenth century and their influence on Finnish and Scandinavian literature (1914-1915), Goethe (1914-1915), late-nineteenth-century Scandinavian literature (1915-1916), Finnish literature (1915-1916), Shakespeare (1916-1918), and post-1809 Finnish literature (1916-1918). In his last year in the professorship he announced lectures on the principal trends of general literary history and the history of nineteenth-century Finnish literature in cultural context. One unique feature of Söderhjelm's teaching was that, in accordance with his mediatory position on the language question, half of it each year was in Swedish, half in Finnish.

Söderhjelm has been characterized as a staunch supporter of positivistic science,[145] and indeed he did believe implicitly in "value-free" scientific inquiry, as this programmatic statement from his 1903 *Euterpe* piece on Gaston Paris makes amply clear:

> I proclaim absolutely and without reservation that doctrine that science has no other goal than truth, truth for truth's sake, without the slightest concern for its possible practical consequences good or bad, fortunate or unfortunate.

In addition to this early empiricism, however, commentators have seen in Söderhjelm's approach the last dying echoes of idealistic aesthetics (Tigerstedt) or a toned-down positivism tied to a critical impressionism (Castrén). Though Söderhjelm never really moved past the methodology he had learned in the 1880s, he did follow current methodological debates, and made a few modifications in his critical approach as a result. His understanding of stylistics took on new depth, for example; he adopted certain of Antoine Albalat's notions; and in 1909 he wrote a favorable piece in the *Neuphilologische Mitteilungen* on Croce's and Vossler's ideas for a new approach to the analysis of literary language.

As a scholar Söderhjelm also practiced rigorous self-criticism. In 1918, when he had virtually left behind him both philological and literary-scientific research, he appraises his work and the choices it was based on in a letter as follows:

> I am not and I have never attempted to be a philosopher, not in the least. For that very reason I may have made a mistake in giving up philology and returning to literature, even though I lack the tinkering mentality and the tenacity required by philology. In fact, I'm probably not much good for anything – unless it is teaching others how to work. I find myself unable to value anything I do, and envy those who like their own books. I don't have a single book that, once it is published, I don't hate to see pulled off the shelf.[146]

The 1910 Appointment

When Eliel Aspelin retired from the chair in aesthetics and modern literature in the spring of 1908, there were only two applicants for his job: Werner Söderhjelm and a man ten years his junior, Yrjö Hirn. The job search once again proved a bitter battle in which differing opinions on the two prongs of the job title played a central part. According to contemporaries, Söderhjelm was a literary historian, while Hirn was an aesthetician whose social-anthropological bent provoked much heated debate.

Upon Aspelin's retirement Söderhjelm was appointed acting professor. At precisely the same time his professorship in Germanic and Romance philology was in the process of being divided into two chairs, and he chose the latter half, Romance philology. Nevertheless, he greatly wanted to exchange his philological job for a literature job, and his chances looked excellent; after all, ten years earlier he had served as peer reviewer for the very same post. In a journal entry for November 15, 1909 the retired Fennophile Aspelin seems to have preferred Söderhjelm over Hirn, despite the latter's "greater originality and profundity" and thus superiority "as a thinker and a scholar," precisely because Söderhjelm was more of a Fennophile (albeit "a dubious one") than

Hirn (179). Aspelin's private vote was also swayed by his conviction that the professorship's central teaching field was literary history, and thus Söderhjelm would better suit the discipline's needs (April 9, 1910; 189).

Aspelin declined the division's offer to write a peer review statement on the two candidates, and so three reviewers were chosen from abroad, a normally problematic decision that was made possible in this case by the fact that neither of the candidates had published anything significant in Finnish. Hirn applied for a lengthy respite period to improve his qualifications, and was granted it, thus extending the appointment process.

In the winter of 1910 the three peer reviewers – Henrik Schück from Uppsala, Otto Sylwan from Gothenburg, and Valdemar Vedel from Copenhagen – submitted a single brief joint statement. In it Söderhjelm was considered qualified for the position due to his aesthetic analytical abilities and his treatment of historical problems, and, because his qualifications exactly matched the chair's teaching fields, ranked first. Hirn's scholarly work was praised as well, but the reviewers found no literary scholarship in his publications, and his theoretical work was more psychological, philosophical, cultural-historical, or ethnological than the "criticoaesthetic analysis" of works of art, which according to the reviewers was "the task of modern aesthetics." Still, due to the high quality of his scholarship and the clear evidence from his work that he possessed a broad familiarity with literature, he was considered qualified for the position.

Despite the unanimous peer review,[147] the appointment process degenerated into an ugly administrative battle. The division of history and linguistics discussed the ranking of the candidates throughout April, on the 6th, the 16th, and the 23rd, with numerous division members defending Hirn against the reviewers. Many took umbrage at the notion that aesthetics is no more than applied art criticism, but there also seems to have been some political maneuvering on the language question. Söderhjelm's strongest backer was professor of Finnish E. N. Setälä, and in general Söderhjelm was more heavily supported by Finnish-speaking professors, Hirn by Swedish-speaking professors.[148] The division vote was even, 5-5, and was decided by Dean Gustafsson's vote for Hirn. Aspelin recorded this decision in his journal as well: "So even now the votes are divided by party: the Svetophiles favor Hirn, the Fennophiles Söderhjelm, even though the latter has always been a lukewarm Finn and has always been on better terms with Swedes than with Finns" (May 8, 1910, 190).

The university senate debated the division recommendation on May 4, with again numerous attacks on the reviewers' conception of aesthetics, and decided in favor of Hirn on May 18. At this level a clear majority found Hirn's aesthetics central to the post, and with a vote of 26-14 he was ranked first.

The outcome of this job search has also been regarded as an implicit policy statement on positivistic science; in this reading a vote for Hirn was a vote for positivism, and his appointment marked the breakthrough of positivism in Finland (Huuhtanen 142-145). It must be remembered, however, that in neither the 1900 nor the 1910 search was any explicit distinction made between trends or methodologies within aesthetics, or even between idealism and positivism. In both cases the major dispute revolved around a choice between literary criticism and the other fields, aesthetics and art history, and in 1910 the language question played a significant if still submerged role. It is true, of course, that the 1900 and 1910 appointments signaled a shift in emphasis from literary and art history (Aspelin in 1900) to aesthetics (Hirn in 1910), and it may well have been that one unrecorded consideration in that shift was a general attitude toward positivism, which in ten years had changed from a cautious and skeptical one to an entirely favorable one.

The New Chair in Literature

Soon after Hirn's appointment, the division of history and linguistics began to consider the possibility of creating a new professorship, one that patently had Söderhjelm's name written on it. In October, 1910, professors Setälä and Ruin proposed the creation of a professorship in Finnish and comparative literary history, evidently with Söderhjelm in mind; but their proposal was voted down 9-6.[149] Yrjö Hirn himself returned to the matter in February of the next year, again proposing the creation of a professorship in Finnish and general literary history, and now explicitly recommending that the job be offered to Söderhjelm.[150] Clearly, Hirn's proposal was something of a conciliatory gesture after the outcome of the job search, but there were strong grounds for the proposal as well. In his written statement Hirn notes that students had long complained about the lack of instruction in the literature of their own country. The professors of Finnish and of Nordic philology were already overburdened with the teaching of practical language skills and language history, and had no time to familiarize themselves with literary texts; and the professor of aesthetics and modern literature was almost exclusively concerned with "European literature."

Hirn argued that, in a country where "the nation's cultured class was made up above all of government officials," academically trained teachers should at least minor in a literary subject. In future it would be impossible for a single chair in literature to handle the resulting teaching load.

Hirn proposed a division of labor between the two literature professors in which only the professor of aesthetics and modern literature would lecture on general theories of art and "beauty" (allmän konst- och skönhetslära) and only the new professor would lecture on Finnish literature, but all other lecture topics would be unrestricted. Thus some overlap between teaching in the "modern literature" of the existing post and in the "comparative literature" of the proposed post would be allowed. Students majoring in aesthetics and modern literature would be required to take courses based around applications of historical, psychological, and aesthetic-critical methods, while those majoring in Finnish and comparative literature would be required to take courses based around applications of philological literary history and stylistics. Coursework in the latter area would be compatible with studies in linguistics, in the former with history, art history, philosophy, and psychology.

In division discussions of this proposal, it became clear that the members would prefer to deal with the creation of a new professorship and offering it to Söderhjelm as two separate issues. A clear majority, thirteen out of fifteen, favored the creation of a new professorship, but only seven favored offering it to Söderhjelm in isolation from other concerns. Since five others were in favor of offering him the job if his professorship in Romance philology were then offered to professor extraordinaire Axel Wallensköld, however, agreement was reached on the main points of Hirn's proposal.

The two professors opposed to the creation of a new professorship were Knut Tallqvist and Ernst Palmén. Tallqvist argued that Söderhjelm could lecture on literature in his current post if he wanted to, and that there were more important chairs than the one proposed, such as in art history or ancient history. In his opinion the division should create new positions not to suit the temporary needs of specific people, but with a view to the long-range development of the university. Palmén agreed substantially with Tallqvist, arguing against placing people before principles and objecting strongly to choosing a successor to Söderhjelm before his position had even become vacant. Edvard Westermarck also read a list of fields that he thought needed a professorship more than Finnish and comparative literature – art history, ancient history, English philology – but voted for the proposal nonetheless.

Fridolf Gustafsson thought that "Finnish literature without a Nordic component would make a pretty thin major," but supported the proposal on personnel grounds: if Söderhjelm had to be placed ahead of Gunnar Castrén, then no doubt Wallensköld could be placed ahead of other possible candidates for the Romance philology job.

The strongest voice in favor of the new chair was E. N. Setälä's, professor of Finnish, who seems to have been thinking as much of what would be best for Söderhjelm as of what would be good for the university. He argued that the Nordic philology instructors could not possibly teach Swedish-language Finnish literature, any more than his Finnish instructors could teach literature in Finnish. The professorship in Finnish, he said, had been considered a philological job from the start (this was, of course, strictly speaking untrue), and even if it were expanded to include literary topics they could only be dealt with linguistically. The university desperately needed a professor who could examine literature as a cultural product, against the backdrop of world literature and its ideological content. As the recent job search had shown, this was not the task of the professor of aesthetics and modern literature, who was to examine the great European literatures in terms of their aesthetic value. In addition, "domestic literature" had never fallen under the purview of aesthetics and modern literature: it had never been taught, and after professor Vasenius had retired it had not even had an advocate in the university.

As an additional consideration in favor of the new post, Setälä cited the wishes of "mother-tongue" teachers, teachers of Finnish, who, he said, had often complained that they needed Finnish literature as part of their training. Finns' all-round education was nowhere as deplorable as in literature, which, he declaimed in idealistic tones, "brings the thought of the times into a focal point, as it were, and then in the most accessible form disseminates that thought into higher and higher spheres."

The university senate discussed the new chair in April, 1911, and voted in favor of Hirn's proposal. Indeed even the senate discussion followed the lines Hirn had laid down in his proposal, though some thought was given to a modified job title. In the end, Hirn's proposed title "inhemsk och allmän litteraturhistoria" was accepted, and the senate also agreed to offer the job to Söderhjelm.

Another argument for the new chair was offered in a letter from acting chancellor Edvard Hjelt to the university chancellor: because the professor of aesthetics and modern literature had to serve as the foremost authority in his field on numerous peer review and other committees dealing with matters related to art and literature, the drain on his time alone was sufficient reason to create the proposed post.

The by-law creating the new professorship in Finnish and comparative literature was signed in Tsarkoye Selo in March 1913, and Söderhjelm was officially appointed on March 22.

The new job title is worth closer attention. During the time when the new chair was being processed, the terms "inhemsk litteratur" and its Finnish translation "kotimainen kirjallisuus" recurred frequently in division debates in the context of two new docentships. Both Viljo Tarkiainen and Arvid Mörne applied for docentships in 1911, and, after long debates, a good deal of severe criticism, and close votes, the division decided in December of 1911 to recommend that both be appointed.[151] Tarkiainen was criticized for the superficiality of his study of the *Kalevala* and what professor Suolahti called his "lack of a finer poetic sense" and Söderhjelm called his "lack of sensitive poetic understanding." Mörne in turn was criticized for the weakness of his dissertation. The senate overturned the division recommendation – on Tarkiainen by a vote of 9-5 and on Mörne by a vote of 11-3[152] – and it was only upon a new application that both were appointed to docentships.

More interesting than the often rather shallow appraisals offered by division members of the candidates' research, however, were the proposals made as to their fields of expertise. Von Bonsdorff's recommendation that Tarkiainen be given a docentship in "literature in Finnish" (in order, he said, that the university gain more teaching power, even if it means "no substantial increase in intellectual capacity") was obviously motivated by the language question. Hirn agreed, though he was willing to accept the title "domestic literature" (inhemsk litteratur). Virtually every member of the division expressed his opinion on the title, with von Bonsdorff and Westermarck preferring "literature in Finnish" (suomenkielinen kirjallisuus), Söderhjelm, Ruin, and Heikel preferring "Finnish literature" (suomalainen kirjallisuus, finsk litteratur, by which they seemed to mean the same thing as "literature in Finnish"), and Lagerborg, Suolahti, Vennola, Krohn, Danielsson-Kalmari, Gustafsson, and Setälä preferring "domestic literature" (kotimainen kirjallisuus, inhemsk litteratur). Setälä, in fact, would have liked Tarkiainen to be appointed "docent in the history of domestic or perhaps better our *own country's* literature." Setälä thought "Finnish literature" a possible alternative, but considered it unadvisable to restrict a docent's teaching field excessively with a job title, especially in the light of Tarkiainen's work on Finnish literature both in Finnish and in Swedish. The division's recommendation ultimately was that Tarkiainen be appointed docent in domestic literature.[153]

A similar debate was waged over Mörne's job title a week later, but with different terms. Here too "inhemsk litteratur" won the day, but "inhemsk litteraturhistoria" (domestic literary history, von Bonsdorff) and "nordisk litteraturhistoria" (Nordic literary history, again by von Bonsdorff, and seconded by Krohn in the event that "inhemsk litteratur" were not approved) were also proposed. Setälä again proposed the title "docent in the history of our own country's literature" and vigorously attacked the notion that Mörne might ever be thought of as a scholar of "Nordic literature."

This job title debate, which was waged shortly after the new chair had been named, is interesting for the light it can shed on how the study of Finnish literature was conceived in the early decades of this century. Part of the division felt strongly that literature written in Finnish should be kept separate from that written in Swedish, and that this discrimination should be made in docent job titles. That Snellman's coinage of a half century before, "inhemsk litteratur," including under the same umbrella literature written in both Finnish and Swedish, should carry the day, was to have far-reaching consequences. In the early 1920s, for example, shifts in the university job structure led to the appointment of Castrén and Tarkiainen as professors of "domestic" literature, and in every Finnish literature appointment since that time the word "kotimainen" or domestic has systematically been preferred over all other alternatives.

Yrjö Hirn

Yrjö Hirn (1870-1952) is one of the brightest international stars in the Finnish humanities this century. He was born in Lappeenranta, where his father taught history in high school and played an active and leading role in local Fennophile politics.[154] The Hirn family later moved to Porvoo, where one of Yrjö Hirn's teachers was the Runeberg enthusiast J. E. Strömborg. After his high school graduation in 1888 he entered the university to study aesthetics and modern literature and philosophy, taking his Master's degree in 1892. He began working in the university library while still a student, and after he completed his studies continued working there, eventually becoming a senior amanuensis, until his appointment to the professorship in 1910.

Though Hirn was eleven years younger than Söderhjelm, both men were part of the transitional generation of the 1880s that threw off the old school's idealistic aesthetics and literary theories. Hirn's career was given a significant direction when he was in England in the 1890s: he spent the spring of 1893, from January to May, in London, researching his 162-page dissertation *Förstudier till en konstfilosofi på psykologisk grundval* ("Prolegomenon for a Philosophy of Art Grounded in Psychology," 1896), and returned to England (both London and Oxford) in 1898 for a two-year stay, during which he prepared a supplementary sequel in English to his Swedish dissertation.

By the 1890s the era of Vischerian aesthetics was fading fast, and evolutionary theories were invading the humanities from the natural sciences. The Finnish pioneer in applying evolutionism to sociology was Hirn's friend Edvard Westermarck, the founder of Finnish sociology, and Westermarck had a great impact on the theoretical alignments of Hirn's own work on art. Both Hirn and Westermarck studied and did research in England, the home of Darwinism, and in 1893 Westermarck introduced Hirn to the relevant academic circles in London, where Richard Wallaschek's ideas among others were to shape his dissertation. In 1896 Westermarck would serve as official faculty opponent at Hirn's dissertation defense, and in 1897 in an article in *Valvoja* entitled "Vertailevien taidetutkimusten metoodeista" ("The Methods of Comparative Art Research"), Hirn would describe his method as "English" (451).

Hirn's dissertation, a psychological and sociological inquiry into the origins of art, was the first in a long series of ethnosociohistorically oriented works he was to write. Proceeding darwinistically through four chapters on "animal art," "feelings and expression," "objectless ecstacy," and "pleasure," Hirn argues that primitive art isn't art at all in the modern sense of the word, but rather part of the practical activities of the primitive society, religious, magical, or military. The dissertation was in fact a kind of introduction to the book Hirn published in London four years later entitled *The Origins of Art* (1900). In both of these works, as in the later popularizing *Det estetiska lifvet* ("The Aesthetic Life," 1913) and *Konsten och den estetiska betraktelsen* ("Art and Aesthetic Reflection," 1937, based on a lecture series he gave at the University of London), Hirn developed the rudiments of his psychosociological method.

Though like these others *The Origins of Art* is an aesthetic rather than a literary-critical study, it will pay us to look briefly at some of its methodological principles, for the same principles inform his literary studies as well. In his discussions of the origins of art Hirn argues that the primitive forms of art grew out of practical needs, such as the expression of ideas, the arousing of sexual interest, the stimulation and maintenance of a working or fighting spirit, or placating the gods; and that aesthetic attitudes are the result of a much later development. In *The Aesthetic Life* he illustrates this process as follows:

> Art historians have shown that the ornamental repertoires of both savage or barbarian and civilized peoples consist largely of combinations of lines derived from the manufacture of objects and mechanically borrowed by more recent technology from older. Thus the lines impressed in the soft clay of a pot by the woven basket that reinforced it during its manufacture are later transformed into ornamental designs. (48)

Similarly Hirn speculates that the origins of the verbal and visual arts lie also in some practical need or necessity "whose effects have shaped our taste and therefore survived in our behavior even after the actual 'necessity' has been removed" (49). For Hirn it was important for us to understand this evolutionary mechanism in the development of art, for it helps to understand the double role of modern art: we no longer believe in its practical usefulness but still project ourselves into it.

> We do not normally look to art for immediate incitements to battle or work, but that increment which art adds to the courage with which we face life does see us through many struggles and troubles. We do not exactly allow artistic works or performances to direct our sexual choices, but dances, poems, pictures, and ornaments do nevertheless directly affect our spiritual life, through attraction and arousal. And while we no longer view art as a source of knowledge, we value the profundity and breadth aesthetic experience lends to our philosophy of life. We rightly insist, therefore, that art never ceases to teach us, or to charm us, or to vitalize us – that it never loses its magic. In part due to these very qualities art has had and continues to have such a central place in human life. (63-64)

Hirn argues that in addition to its implementational value art has always had value in itself, a value arising out of a human "expressive drive." Implementation generated the framework within which the expressive drive operated (77-78). In speaking of this human drive toward self-expression Hirn characterizes art as a transmitter of feelings: whether the artwork is highly controlled intellectually and thus capable of developing into "an explanation of the world or of life," or the mere spontaneous expression of the artist's individual experience, emotional elements remain important to art. They separate art from other human behaviors, and constitute the only common denominator joining all forms of art (91-92).

There is a certain duality to Hirn's aesthetic thought.[155] Despite his evolutionary model and its attendant impulse to rationalistic, scientistic explanation, he also recognized the existence of what he called art's "miracle." He sought to combine the psychosociological examination of art's marginal conditions with an inquiry into its purely aesthetic "miracle." Unlike the other Finnish positivists of the day, such as Vasenius, Hirn perceived this duality in his own work and sought to integrate it into a coherent whole, rather than believing blithely in the omnipotence of positivism and remaining blind to the vestiges of idealism in his thought.

Still, Hirn's approach to art is fundamentally darwinistic, grounded in the study of origins and developments. The primacy Hirn assigned practical needs in the origins or art is no longer beyond dispute, and his theories of the transmission of emotion have virtually been discarded. But his art theories remain interesting in many respects. Art's dual life, for example, its participation in both practical needs and self-justificatory expression, recalls the conception of art's various and historically contingent functions advanced by Jan Mukarovsky in the 1930s.

Upon his appointment to the docentship in aesthetics and modern literature in 1898, Hirn mainly lectured for the first few years on the early history of art from an aesthetic and cultural-historical perspective and related theoretical concerns: the development of dramatic art and the history of aesthetics (1899-1900), "the beauty-sense, the nature-sense, the comedy-sense, and their development" (1900-1901), John Ruskin and the manifestations of French classicism in poetry and art theory (1901-1902), forms of religious art (1905-1906), and general aesthetics (1906-1907). In addition he lectured in the spring of 1901 on the Pre-Rafaelites (Brown, Millais, Hunt) and the decorative art industry, which was best-known in Finland through the Jugend style (Rantavaara 1:154).

In the first years of the century Hirn served for a year as chairman (1901-1902) of the newly founded and short-lived Students' Aesthetic Association (1901-1903).

In the summer of 1902 Hirn received the three-year Rosenberg fellowship, which once again made it possible for him to travel extensively in Europe doing research. His aim on this trip was to gather material on medieval Catholic thought and painting and the legendary and narrative traditions associated with them. He traveled through Germany to Switzerland, from there to Italy, where he spent four months in Florence, haunting the art centers and art museums both there and elsewhere in Italy. After a brief visit to Finland in the summer of 1903, he continued his journey to Paris, established

Yrjö Hirn, pencil sketch by Magnus Enckell in the early 1890s. (M)

a base there and made side trips to Spain and the Netherlands, not returning home until the summer of 1905. After another funded trip to London and Oxford in 1908, Hirn was ready to publish a treatise on Catholic ecclesiastical art and poetry entitled *Det heliga skrinet* ("The Sacred Shrine," 1909). The book is in two parts, the first (3-160) dealing with the "holy casket" and the second (163-458) dealing with a symbolic holy casket, i.e., the Virgin Mary, and related forms of worship. The first part ranges widely through architecture, art handicrafts, and sacramental pantomimes used in the Mass and Communion; the second covers sculpture, painting, and poetry. Hirn himself was not religious, and his perspective throughout is detached, the view from outside of a social observer. In Finland Hirn had campaigned for the disestablishment of the Lutheran Church, and when *Det heliga skrinet* appeared in English translation several reviewers accused him of anti-Catholicism (Rantavaara 1:210-211).

Hirn made his last long trip abroad in the mid-1910s, traveling first to Switzerland to do research on Rousseau and children's culture; then, upon the outbreak of the First World War, continuing on to London, and returning in May, 1915 to Finland. The Rousseau book never came out (Hirn did publish several articles on him), but the book on children's games, *Barnlek*, appeared speedily (1916), with chapters on toys, games, children's dances, children's songs, marionette and Punch-and-Judy shows, shadow plays, and the circus. In his preface to the Finnish edition (1918) Hirn states his purpose in the book:

> In each chapter I explore some unique aspect of the aesthetic life expressed in the games and dreams of small children. I wanted to draw out some of the poetry, or rather the romance, of that age, which is still so alien to the intellectual world of adulthood, but which receives so many indirect stimuli therefrom. In particular I hoped to show that the playful expressions of the childish imagination are, despite their gay and apparently mindless nature, closely tied to the trials and tribulations of life and to thoughts about its grim reality. (5)

Hirn's aesthetic works rapidly brought him international fame. *The Origins of Art* appeared in Swedish translation in 1902, in German in 1904, and in Russian in 1923. *Det heliga skrinet* appeared in English translation under the title of *The Sacred Shrine* in 1912 and in partial French translation; *Barnlek* appeared in French in 1926 and in Italian in 1929. A great number were also translated into Finnish.

As is clear from the above, Hirn's area of expertise was broad, though it remained roughly within the field of aesthetics and the sociology of art. In addition to this work on the psychology and sociology of marginal or borderline art, in 1949 he published a book on nineteenth-century theater life in Finland (*Teatrar och teaterstrider i 1800-talets Finland* ["Theaters and Their Struggles in Nineteenth-Century Finland"]).

But he also published on literary subjects, in three areas: eighteenth-century English and French literature, the history of specific themes and motifs, and Runeberg. The eighteenth century was a literary period Hirn felt close to. Indeed it has been said that he found in the Enlightenment the kind of "clarity of thought and humanity of feeling" that he himself possessed.[156] His books on this period include *Diderot* (1917), *Swift* (1918), *Dr Johnson och [and] James Boswell* (1922), and *Beaumarchais* (1931), as well as the articles on the literature and cultural history of the period collected in *Episoder* ("Episodes," 1918). These titles alone suggest Hirn's interest in borderline cases, literary and social marginality. His first books on literary topics were popular biographies that appeared in a series entitled "Notable Men." He himself served on the editorial board for this series, and wrote *Diderot* and *Swift* for it. He had done the research for the former back in 1911, when he had given a 25-hour lecture course on Diderot, and the Swift book took shape in similar fashion, based on lectures he gave in the spring of 1916 (Rantavaara 2:57, 70). In these two books Hirn examines the eighteenth century through the lens of a darwinistic cultural historian, who yet shows a sensitivity for the subtle aesthetic nuance. Though Diderot, for example, whom Hirn greatly admired, anticipated the industrial-art mentality of Ruskin and Morris, in his depictions of technical tools

> he most probably did not yet understand how an evolutionary stage can breathe a charming and, if you will, even philosophical life into the history of the simplest tool. But while he lacked the evolutionary scientist's ability to perceive the connections between primitive tools and complex machines, he still had the inbred fascination of the craftman's son with everything that moves on wheels, belts, and levers. (50)

In his discussion of *Gulliver's Travels*, too, he juxtaposes Swift's work with the present, the period of the Great War: in his third and fourth books, for example, Swift failed to trace the development of contemporary society far enough to understand fully "how compatible the highest possible scientific-technological sophistication is with the lowest possible moral and ethical degradation." Hirn's own time was a perfect example of a technological Laputa whose "much-acclaimed victories" had been all too evident over the past years (109).

Hirn's special interest in borderline cases is clear in his praise for Diderot's sympathetic treatment of the "mechanical arts" (by which he means handicrafts) in the *Dictionary*. Here too Diderot anticipates later developments.

> I may be reading too much into this, but it seems to me that Diderot may have sensed something of that great and wonderful (though according to recent studies too impractical) notion of art's connection with and improvement and liberation of handicrafts, which was the foundation for John Ruskin's and William Morris's enthusiastic labors. (49)

Again, his admiration for and interest in medial forms, gray areas, comes out in his claim that Diderot's magnum opus was his correspondence: the letters reveal the writer's personality more clearly than perhaps any other work in world literature, with the possible exception of Boswell's life of Johnson.

Hirn distances himself from the contemporary biography boom by saying early on in the Diderot book that Diderot's life was not, in the strictest sense of the word, "interesting." Only by delving deeply into his works does his life take on a deeper interest. Hirn interprets literary works as the reflections of their author's world view. He characterizes Swift as a masterful portrayer of horror and disgust, for example, and suggests that Gulliver's adventures in the court ladies' dressing room reflects "the biological aspects of Swift's pessimism" (106). And while *Gulliver's Travels* dazzles Hirn for the "genius of its farcical eccentricity," in relation to Swift's life it is a heart-rending document of human betrayal (109).

In the articles collected in *Episodes* Hirn sought to flesh out, through small but revealing vignettes, a picture of the "fundamental quality of eighteenth-century life," the period that, as he says, "showed the way for the youth of the 1880s." The book's preface, from which these quotes are taken, is dated August 1918, in a newly independent Finland stained with the bloody events of the Civil War and the misery of the imprisoned Red soldiers thereafter. The preface tonally reflects the disappointment with the "present moment" of a liberal whose social roots were in the 1880s:

> It has been refreshing to pause to study the century whose grand ideas showed the way for the youth of the 1880s and whose rewarding philosophy of life, despite all its failings, still offers such a striking contrast to the new ideals. It is tempting to let one's imagination meander down the winding paths of the past, for even if one lingers on the road, one always finds one's way back home, to the present moment. (iii-iv)

Episodes is Hirn's first extended exploration of that field of marginal art that was to concern him for over a decade. As its title suggests, the book presents episodes in seventeenth- and early-eighteenth-century European cultural history: the popular mimoplastic performances (a chapter beginning with Madame de Staël's visit to Finland), the traditional ritual play of the Vevey winegrowers, Rousseau's Thones journey from the perspective of romantic love, Diderot's paradox of the actor. In this last Hirn also took the opportunity to air his conception of art:

> One begins to read the piece with the idea of learning something about the theater and drama; but one soon discovers that reading leads one from this specific case to the more general problems of aesthetic life: the paradox of the actor points to the paradox of art, which must evoke an immediate experience of life in the most consciously controlled form; which must appear to be nature without ever becoming mere nature; which must be created with immense effort and often much suffering, but which must evoke feelings of light and liberated pleasure; which must be at once personal and universal. (319)

Hirn's books of the 1920s – *Eremiter och pilgrimer* ("Hermits and Pilgrims," 1924), a study of natural and religious feelings in the light of ancient literary documents; *Den gamla postvagnen* ("The Old Stagecoach," 1926), a book about travelogues of the romantic era; and *Ön i världshavet* ("Island in the Ocean," 1928), which explores the history and shifting popularity of the motif of castaway/island life – form a kind of trilogy of theme and motif history, in which Hirn is again interested in marginal literature as much as he is in literary classics. Later books in the same general realm would include his *Lärt folk och landstrykare* ("Learned Folk and the Vagrant," 1939), a book about the Finnish mystique of Karelia, a turn-of-the-century ideology that Hirn called (in a coinage that has become part of the language) Karelianism; and *Goda vildar och ädla rövare* (1941), which explores the motifs of the honorable thief, the noble savage, and the rogue. Hirn ties the phenomena he discusses to a broad cultural and social context, and in modern fashion explains fluctuations in the popularity of his motifs in terms of that context.

Hirn himself groups the three books of the twenties together in his preface to *Island in the Ocean*, saying that in all three he is concerned with individual cases and is not interested in formulating a general explanatory model or theory of the travelers or hermits or castaways he explores, for "only the detailed study of individual cases without preconceptions can produce reliable results" (7). Nevertheless, Hirn's psychosociological vision informs all three books: the travelers, hermits, and castaways are joined by a common human thread, the desire to retire into solitude, *Wanderlust*, which the community has occasionally legitimated (mendicant friars), and a latent critique of "cultivated society" made possible by special circumstances. The psychological and the sociological explanations are complementary: for example, the solitude of a deserted island appeals to basic human needs, the satisfaction of which only became possible in the age of exploration, and the popular depiction of which only became literarily feasible in the age of romanticism.

Hirn's research interest in Finnish literature was Runeberg. His first and, compared with other Finnish Runeberg criticism of the time, most interesting book on Runeberg was grounded in his concern with cultural sociology. In the 491-page *Runebergskulten* (1935) Hirn explores the forms and history of the Runeberg cult, which had first touched him personally during his school days in Porvoo, where the Runeberg enthusiast Strömborg had been one of his teachers. Hirn analyzes the Runeberg cult from the outside, critically, without being swept up, as the earlier generation had, in the current of the cult. He rehearses contemporary opinion of Runeberg, honors and criticism, later annual commemorative celebrations, and the cults and legends surrounding various places associated with Runeberg. Hirn explains the formation of the cult by reference to favorable circumstances: a national hero was needed in the formative period of Finnish nationalism and relative political autonomy. Thus also the obsessive need to deny Runeberg's foreign influences. Hirn's approach is shaped partly by his knowledge and experience of religious ritual, of course, and while in his preface he gives the Runeberg cult his wholehearted (if perhaps also politic) approval, the dry, detached, "sociological" tone of his book was novel in connection with Runeberg:

> Needless to say, we see nothing wrong with the Runeberg cult; on the contrary, we consider it an honor to Finland that, in the course of the last century, specifically beginning with the publication of Runeberg's first poems, a cult of this sort took on the magnitude and significance that it did, that it became very much the leading patriotic cult. This is why it seemed worthwhile to collect and collate information about the various forms of that homage which the Finnish people, with perhaps a greater piety than any other people, have given their poet. In presenting our observations we have therefore thought it appropriate to retain the perspective employed by the ecclesiastical reviewer [who had attacked the religious forms Runeberg worship was taking in the 1880s], i.e., the same analytical approach one would use to study the cult of a hero or a saint. (3)

Methodologically Hirn's book is close to the sociology of literary taste, a method that was associated with Levin L. Schücking between the wars. On the other hand it also obviously arises out of Hirn's long-standing concern with the sociology of religion, which had produced *The Sacred Shrine*. *The Runeberg Cult* remains one of the most interesting books ever written in Finland on a changing writer-image.

In *Runeberggestalten* ("The Figure of Runeberg," 1942), a collection of previously published articles on Runeberg which Hirn published after his retirement, the main concern is the mystical side of Runeberg's work as a poet. He alludes to the difficulty some earlier critics had had in understanding the significance of powerful emotion and certain metaphysical statements in Runeberg's poetry. He notes that Strömborg, for example, systematically ignored certain of Runeberg's expressed opinions, no doubt because they seemed "otherworldly" to him. Otherworldly is in fact Estlander's word; but Estlander had shown that the otherworldliness had its counterparts in the religious meditations published after Runeberg's death. Hirn continues:

> But there is more. We believe that the same otherworldliness, or rather, as we should prefer to say, the same mystical conception of nature

> appears not only in Runeberg's poetry and religious thought, but in his secular writing as well, and indeed is central to its essence. (199)

This reflects Hirn's own personal predilection. He was attracted to the notion of a Great Moment, which was not institutionally religious but rather mystical in content. Even in the early work on the origins of art he had stressed the momentary event, the sudden experience in which practical work was transformed in the mind into art, in which the painted pot in the primitive potter's hands suddenly appeared beautiful.

In the last section of the book Hirn analyzes Runeberg's poetic conception and treatment of individual freedom and its limitations – a recurring theme of Runeberg criticism, running from Snellman's first essays to contemporary Finnish Runeberg critics.

While Hirn was a docent his teaching was focused on aesthetics. As Aspelin's retirement drew near, he (Hirn) began to shift more in the direction of literature, lecturing on the history of literature and the theater (1908-1909), Rousseau (1909-1910), and other literary subjects, until his appointment to the professorship shifted the focus of his teaching almost entirely into literature. In addition to aesthetics, in the 1910s Hirn lectured on eighteenth-century French literature (1910-1911), Voltaire, Holberg's comedies (1911-1912), the evolution of poetic forms, the use of classical conceptions of tragedy in modern literature (1912-1913), Hebbel (1913-1914), eighteenth-century English literature, the French classical drama (1915-1916), eighteenth-century English literature (1916-1917), eighteenth-century English romanticism (1917-1918), late-eighteenth-century English literature, and the plays of Grillparzer (1918-1919). This fairly narrow focus on aesthetics and eighteenth-century English and French literature was made possible by the substantial increase in teaching staff, both the new professorship and several new docents.

Hirn retired from the professorship in aesthetics and modern literature in 1937. Between the world wars he played an active extracurricular role in Finnish cultural life, in areas and ways congenial to his liberalism. His Anglophilia also offered a significant counterbalance to the increasing influence of German art and thought in Finnish cultural and academic circles. Hirn collaborated with the Otava Publishing House in commissioning translations of world literature and a massive Finnish history of world literature (which latter project, as Hannu Riikonen documents, was only partially realized [26-30]). Hirn's cultural-sociological theories and freethinking attitudes fell once again on fertile soil when young members of the Finnish left, which was beginning to recover from the Civil War, became his students. Though he avoided open battles over cultural politics, Hirn acted as a mediator of the liberal and democratic values that had been vital in Finnish literary criticism since Cygnaeus's time, ushering the nineteenth century into the twentieth.

Critics as Library Staff

Around the turn of the century the university library became a veritable powerhouse of significant literary-historical research. The library had on its payroll numerous scholars with research interests in literature, some of whom later went on to make careers for themselves specifically as literary critics. In addition to the long-time head librarian, Wilhelm Bolin, and Valfrid Vasenius, who worked in the library from 1871 to 1901, the following critics all worked at least at some period in their life in the library: Arvid Hultin (1877-1926), Arvid Mörne (1895-1896), Alexis von Kraemer (1897-1899), Gunnar Castrén (1897-1907), Yrjö Hirn (1890-1910), Rolf Lagerborg (1897-1907), a philosopher who was interested in literature, and who was one of the first to introduce psychoanalysis to Finland,[157] Emil Hasselblatt (1907-1943), Olaf Homén (1911-1919), and Alexander Boldt (1894, 1902-1940), a library amanuensis and poet.

It became a tradition for librarians to gather in the so-called "literary hall" in off hours to discuss literature and the theater, which led to the development of what Gunnar Castrén recalls as a "library spirit" (13-15). Since librarians only worked three hours a day (younger librarians even less), a job in the library left plenty of time for scholarly and literary pursuits; indeed the *Euterpe* group was intellectually based in the library.[158]

At the turn of the century the library tradition that had grown out of older scholarship in the history of learning and literature was revitalized above all in the work of Arvid Hultin (1855-1935). Hultin was the son of an associate pastor named Alexander Hultin, and had grown up in the county of Helsinki.[159] He went to school in Porvoo, and in the 1920s was to write the history of his school; after graduation he entered the University of Helsinki, where he studied general history and aesthetics and modern literature, taking his M.A. in 1878. During Hultin's years at the university ideological Svetophilia was on the rise, and he imbibed a good deal of it, both from the Uusimaa Students' Club and from his professor, Estlander. In his later scholarly work Hultin was largely interested in the history of Swedish-language literature and culture in Finland.

Beginning in 1879 Hultin worked as a library amanuensis, but he also taught school and worked as a reporter for the Svetophile daily *Nya Pressen* from 1883-1900. From 1903 to his retirement in 1926, he served as assistant head librarian and director of the Fennica collection, which comprised Finnish literature (Nohrström 135-163). While editing an index of Swedish-language literature in Finland up to and including 1885 (*Den svenska skönlitteraturen i Finland till och med år 1885*, 1885) and reorganizing the Fennica collection on the Uppsala model, thematically and alphabetically, Hultin thoroughly familiarized himself with earlier Finnish literature written in Swedish.

At Estlander's suggestion the Swedish Literature Society had started a series in 1893 entitled Finlands svenska vitterhet (Finnish Literature in Swedish), which was designed to reprint works in Swedish from the era of Swedish rule as well as biographical studies of their authors. The series was to be divided into three periods: (1) the Great Power era, (2) the Age of Freedom and Gustaf III's reign, and (3) the Franzén era, concluding with the university's move to Helsinki. The plan was also to publish a literary history of each period, characterizing its general features, describing the work of minor writers, and printing selections from their works. Estlander insisted that major-author monographs and reprints be aimed at the contemporary reader, and thus be equipped with biographical and other useful historical information; and second, that only texts that were significant in terms of the artistic standards

The reading room of the University of Helsinki library. (M)

of their day be included.[160] The first volume to appear in the series was M. G. Schybergson's edition of *Jakob Tengströms vittra skrifter i urval* ("Selected Literary Writings of Jakob Tengström," 1899), which also included a 146-page biography of Tengström. The second was Ernst Lagus's *Mikael Choraei valda dikter* ("Selected Poems by Michael Choraeus," 1901), with a 190-page biography.

The main responsibility for editing this series rested with Hultin, who had served as the Society's secretary since 1895 (and would remain its secretary until 1935, forty years all told). In addition to editing the Society's publications he also published several of his own studies in the Society's publication series.

Hultin published his first article in a Swedish Literature Society series in 1889, on Anders Oskar Roos ("En finländsk dramatiker på 1830-talet" ["A Finnish Playwright in the 1830s."]). In the articles he wrote in the early 1890s on late-seventeenth-century Finnish literature, on professor of eloquence Daniel Achrelius, and Michael Renner, Hultin found his true scholarly home, Finnish literature and cultural history under Swedish rule. In his later monographs on writers – his dissertation *Torsten Rudeen* (1902), *Jacob Frese* (1902), and *Gustaf Filip Creutz* (1913) – and his literary histories – *Den svenska vitterheten i Finland under stormaktstiden 1640-1720* ("Swedish Literature in Finland During the Great Power Era, 1640-1720," 1920), *Finlands litteratur under frihetstiden* ("Finnish Literature During the Age of Freedom," 1906), and *Det ekonomiska tidevarvet i Finlands litteraturhistoria* ("The Economic Period in the History of Finnish Literature," 1910) – Hultin explored the earliest literature of Finland at great length, almost 2000 pages. He also sought to define the Finnish identity of writers who lived in Finland and wrote in Swedish. According to Hultin, for example (drawing on Cygnaeus's work from mid-century), Frese's poems reflect a unique combination of ardor and melancholy (innerlighet och vemod), like an emergent symptomatology of the musical quality of the later poets Creutz and Franzén (119).

The above-mentioned works were supplemented in 1918 with a collection of articles published since 1914 (a book with a print run of only nine copies), *Litteraturstudier i Åbo under svenska tiden 1640-1809* ("Literary Studies in Turku During the Swedish Era 1640-1809"). This was a comprehensive history of literary learning at the University of Turku. Indeed Hultin's earlier works had combined the history of learning and the history of literature on a broad scale as well, and like this one they too had dealt extensively with the evolution of various branches of learning at the university. According to its subtitle, for example, Hultin's dissertation on Rudeen was "an addendum to Carolinian literature and history of learning"; and the sections of his literary histories dealing with the eighteenth century focused on Linné's student Johan Brovallius, who in 1737 had been appointed professor of physics in Turku, as well as the internationally known natural scientists Per Kalm and Pehr Gadd.

Hultin had only originally planned to write his literary history through 1720, but when Ernst Lagus decided not to do his segment on the age of freedom, Hultin researched and wrote up that segment as well. The same thing happened with the monograph on Creutz: Estlander had promised to write it, and when he gave up on the project it fell to the Society's secretary.

Because the Finnish Literature in Swedish series was not shaping up as planned, upon Estlander's death in 1911 the Society decided that it would be concluded after the publication of Castrén's edition of Franzén's Finnish-period poems and Ernst Lagus's edition of minor poetry from 1721 to 1809, both of which were in progress at the time (Steinby 133). Neither in fact was ever completed, and the concluding volume in the series remained Hultin's volume on Creutz.

In his reading of Creutz, which he attached to a new edition of Creutz's poems, Hultin pays particular attention to the poet's influences, and his approach throughout is closer to cultural history than literary criticism. His research in difficult-to-find primary source material is impressive in its exhaustiveness and also in the intelligence of its analysis, and he ends up providing a wonderfully clear and complete cultural and biographical context for Creutz's literary work. His analysis of individual literary texts is more superficial, less well grounded. He devotes the lion's share of the volume to the biography (380 pages), leaving 100 pages for all of Creutz's published poetry.

Hultin's position on the language question was exaggeratedly Svetophile. Toward the end of the nineteenth century the Fennophile and Svetophile camps had grown slightly closer, but after 1905 they began to become more polarized again, and particularly beginning in the 1910s the Svetophile camp began to hone their position terminologically: a person from Finland, regardless of language, was to be called a "Finlander," while the term "Finn" was to be reserved for those whose native language was Finnish.[161] It was, quite obviously, a matter of some consequence for the writing of literary history what stance one took on the language question. In 1914, for example, Gunnar Castrén took an antiseparatist stance, calling for connections with Fenno-

Board meeting of the Swedish Literature Society in Finland, 1910. From the right: Hugo Pipping, Fridolf Gustafsson, Sigfrid Kullhem, Oskar Hultman, Petrus Nordmann, Reinhold Hausen, R. F. von Willebrand, M. G. Schybergson, Arvid Hultin, Elis Lagerblad, Ernst Lagus, A. O. Freudenthal, Walter Runeberg and Carl Synnerberg. (SLF)

phone culture as in the best interests of the Svetophone minority. Hirn and Söderhjelm likewise avoided being too closely associated with the Svetophile camp.

Hultin's Svetophilia was evident in his pointed 1917 review of a literary history by Ruth Hedvall, for example, which did not in his opinion sufficiently take into consideration the work of the Swedish Literature Society, the *Finsk Tidskrift*, and Estlander – all of which since the 1870s had insistently campaigned for a recognition of the autonomy of Swedish-language culture in Finland. In addition to the polarization of the language question, Hultin's review and the debate it occasioned[162] reflect significantly divergent conceptions of literature. Hultin and Hedvall are probably the clearest representatives of the schism among Swedish-speaking scholars in the early years of this century, specifically in their conceptions of how literary-historical data was to be defined and how it was to be analyzed. Hultin was closer to cultural history in his insistence that the literary scholar never restrict his or her focus to *belles lettres* but encompass other writing as well, and that, like nonfictional writing, *belles lettres* too be examined in terms of the history of ideas and influences. Hedvall on the other hand favored restricting literary studies to *belles lettres*. In his review Hultin noted that, contrary to established practice, Hedvall had treated literary history as if it consisted purely of *belles lettres*, and that this was an inappropriate assumption to make about Finnish literature written in Swedish:

> The history of Finnish literature written in Swedish poses particular difficulties to the scholar who focuses only on *belles lettres*, for there are whole periods in which no *belles lettres* were even written, but which were often rich in cultural or scholarly writings. This is especially true of the eighteenth century. (458)

Olaf Homén (1879-1945) was the son of a high government official who graduated from the elite Helsinki high school Nya Svenska Läroverket in 1897 and began his studies at the University of Helsinki the same year.[163] Soon after he entered the university he went to work in the library, and after his graduation continued to work there full-time until his appointment (after the Swedish scholar Fredrik Böök declined the job, which was first offered to him) in 1919 as professor of the history of Swedish literature (svensk litteraturens historia) at the newly founded Åbo Akademi. In the early 1920s the job title was changed to conform with Swedish practice: the new title read "litteraturhistoria med poetik" (literary history and poetics).

As was the case with so many Swedish-speaking Finnish humanists in this period, Homén's intellectual development was strongly shaped by his association with *Euterpe* magazine. Homén was the magazine's theater critic, and his pointedly analytical reviews had an immense impact on the evolution of the Finnish theater in the early decades of this century. Homén's position on the language question emphasized the cultural heritage of Swedish-speaking Finns, and this concern runs all through his theater criticism. Since most of the actors performing on the main Swedish-speaking stage at the Svenska Teatern were from Sweden, Homén took up the cause of native Swedish-speaking Finnish theatrical culture:

> The question of language is neither more nor less than a question of the force of habit. There is no theoretical or aesthetic reason why one should speak as they do in Stockholm rather than as they do in Helsinki. A day may come, however, when native stage Swedish will become an important weapon in the cultural battles quite rightly waged by a linguistic group for its very survival.[164]

During the first two decades of this century Homén was also an influential essayist and book reviewer. In his survey of the most recent Swedish-speaking Finnish writers in *De nya författarna* ("Recent Writers," 1915), he also coined the word "dagdrivare" ("idlers," now in common literary-critical usage) to describe the generation of writers whose work was full of bleakness and *Weltschmerz*.

For a brief time around the turn of the century Homén had studied in Germany – in the preface to his dissertation he thanked Karl Vossler for his suggestions – and France, and beginning in 1907 he spent long periods in Paris researching his dissertation on French literary classicism. The 421-page dissertation that he defended in 1914, like its shorter sequel in 1916, attempted a synthesis of the French world view and conception of art in the period. In form the dissertation resembles a collection of articles – ten untitled chapters without transitions – but this form does not mean a loss of scholarly rigor.

Olaf Homén (HKM)

Homén's notes make it clear that he was abreast of all the latest critical developments in French literary history, both in France and Northern Europe. Drawing on classical moral philosophy and literary works, Homén offers a definition of the period's conception of humanity, which is grounded, he argues, in an emphasis on the social at the expense of the individual (l'honnête homme).

After his appointment to a docentship in modern literature (nyare litteratur) in 1917, Homén taught a course on Grillparzer's plays (1917).

Homén's research on French classicism was originally intended to be mere groundwork for a massive study of Voltaire, which, apparently due to excessive self-criticism, never appeared. To qualify himself for the professorship at Åbo Akademi he wrote the first volume of a biography of *Gabriel Lagus* (1921), but the second volume to this never appeared either. Throughout his professorial tenure he worked on a poetics that still did not satisfy him at his death, and it was edited for posthumous publication by his successor, Erik Ekelund, under the title *Poetik* (1953). Had it appeared in its own era it might, despite its rather fragmentary theoretical underpinnings, have proved influential in the methodological move toward an exclusive focus on the text.

At the turn of the century Alexander Boldt (1873-1956) wrote his dissertation on what various Swedish and Swedish-speaking Finnish writers' depic-

Olaf Homén, cartoon from the *Lucifer* album.

tions of childhood and adolescence tell us about their personalities (*Studier öfver en del barndomen och ynglingaåldern berörande drag i svenska och svensk-finska skalders diktning* ["Studies of the Treatment of Childhood and Adolescence in the Poetry of Some Swedish and Swedish-speaking Finnish Poets"], 1902).[165] In the first part of the dissertation (1-165) Boldt uses a peculiar combination of minibiographies and extensive textual sampling to cover fourteen writers from Mrs. Brenner and Jacob Frese to Leopold and Franzén. In the second part (169-204) he ponders the concepts of "childhood" and "adolescence" and once again cites lengthy passages from the writers in order to make "excursions" into specific problems. The dissertation is essayistic in style and fails to reach any conclusions or offer any results, and while it is concerned with exclusively literary material Boldt never quite makes clear whether he is doing literary history or a primitive kind of developmental psychology. Vasenius, who served as official faculty opponent, criticized the dissertation for its lack of aesthetic evaluations, but – with an appeal to "a similar case," i.e. Gunnar Castrén's dissertation, for which he had also served as faculty opponent – he did recommend approval.[166] The division did not agree that Boldt's and Castrén's dissertations were comparable, however, and Söderhjelm even wrote a statement holding up Boldt's dissertation as a cautionary example of what "arbitrary subjectivity" in the choice of topic and methodology can lead to. The division failed Boldt's dissertation, with only Vasenius voting for its approval. Boldt later published Erik Gustaf Geijer's biography and a selection of his works in three thin volumes (1909-1910).

Emil Hasselblatt (1874-1954) studied literature first in Finland, then in other Nordic capitals, and then became one of the first Finns to study library science, in Leipzig.[167] Around the turn of the century he was an active book and theater reviewer for several Helsinki periodicals, and in 1907 he went to work full-time in the library. Though Hasselblatt never wrote any of the various dissertations he was planning on Levertin, Kielland, or collisions between national and Scandinavian trends in nineteenth-century Norwegian literature, he was the foremost expert on Norwegian literature in early-twentieth-century Finland. He collected a group of his literary pieces in *Dikt och diktare* ("Poem and Poets," 1918).

Since the bulk of Arvid Mörne's and Alexis von Kraemer's literary criticism was written after they left the library, and Gunnar Castrén's (like Mörne's) scholarly work was tied through a docentship to the department even while he was working at the library, these three scholars will be discussed below.

Improving the Department Faculty

Teaching in aesthetics and modern literature throughout almost the entire 1890s rested entirely on the shoulders of Estlander and Aspelin (Tikkanen only taught art history). Toward the end of the century graduate studies in literary criticism began to expand, however, as is clear from the marked increase in published dissertations. An influential figure behind this expansion was Werner Söderhjelm, whose work in philology since 1894 had directed students toward modern European literature. Beginning in the new century, and particularly in 1910 with Aspelin's retirement, as the fields covered by the professorship in aesthetics and modern literature came increasingly into question, the teaching staff in the department was rapidly expanded with new docents: J. O. E. Donner in 1897, Yrjö Hirn in 1898, Gunnar Castrén in 1907, Emil Zilliacus in 1909, Jalmari Hahl also in 1909, the aesthetician K. S. Laurila in 1910, Viljo Tarkiainen in 1913, Arvid Mörne also in 1913, and Olaf Homén in 1917. After Finland gained its independence, the two literature departments had a total teaching staff of two professors and seven docents, which guaranteed a scholarly base for the expansion of literary studies outside Helsinki to the two new universities founded around this time, Åbo Akademi and the University of Turku.

The first of the new docents to defend his dissertation was J. O. E. Donner (1864-1942), who was born in Kokkola, graduated from high school in 1883, and completed his Master's degree in aesthetics and modern literature in 1889.[168] After briefly teaching German in a Pori school, he continued his postgraduate studies in language and literature in Berlin and Paris in 1891 and 1892, and in the years following in London.

Donner's literary criticism deals primarily with nineteenth-century German and English literature, with an attempt at methodological modernity: he combines comparative influence studies with worldview research.

He got his dissertation topic in Berlin: many German scholars, particularly Wilhelm Dilthey in his studies of Novalis (1865) and Schleiermacher (1870), had alluded to the importance of Goethe's *Wilhelm Meister* for the German romantics, and in his 211-page dissertation *Der Einfluss Wilhelm Meisters auf den Roman der Romantiker* Donner explores this influence in terms of the forms the "romantic imagination" gives the moods, events, and destinies of *Wilhelm Meister* in later romantic novels (ii). Donner assigns central importance to imitations of the *Bildungsroman* form, and so focuses on seven German writers (Ludwig Tieck, Friedrich Schlegel, Dorothea Schlegel, Novalis, Clemens Brentano, Joseph von Eichendorff, and Karl Immermann) whose novels of the period between 1798 and 1836 he reads as influenced by *Wilhelm Meister*. In his lengthy introduction, for which Erich Schmidt had made suggestions in Berlin, Donner characterizes the reception of *Wilhelm Meister* and its emergence as the German romantics' favorite; in later chapters he then examines individual novels in terms of the elements they have borrowed from Goethe's book. These include the "Nichtsthun" of the hero, an occasional sensuality, a German-Italian elective affinity, and the recurrence of certain characters (Philine, Mignon). In all the dissertation is a fairly unpretentious and unproblematic influence study concerned mainly with similarities; divergences do not seem to interest Donner much.

After 1895 the University of Helsinki no longer required a separate dissertation for docentships,[169] but Donner's 135-page monograph *Lord Byrons Weltanschauung* (1897) was clearly intended to demonstrate his qualifications for a docentship. Donner argues that we have adequate knowledge about Byron's life, but inadequate knowledge about his world view, and so, taking world view to mean above all religious thought, attempts to determine Byron's "view of the great world enigma" (30). According to Donner Byron deliberately gave out conflicting information on his world view, which in turn led to the unsatisfactory state of Byron world view criticism to date: he has been seen as either uncommitted and unreligious or as an open atheist (16). Based on close readings of biographical facts, Donner argues that Byron was instead deeply if nondenominationally religious. Despite his rejection of all churches, he had a "profound respect for divinity" (30). Donner largely takes his evidence from the poetry. Contrary to previous critical opinion, Donner claims that Byron was a pantheist poet, and shows at some length that Byron's *Weltschmerz* does not make his world view a negative one. It simply forces the poet to take refuge in nature, which leads him to peace and oneness with the divine. Byron's *Weltanschauung* is in effect the product of his *Weltschmerz*.

Donner's scholarly output was small, but his appointment to a docentship in aesthetics and modern literature in 1897 (despite the fact that he was frequently on leave of absence) meant a substantial increase in the department's teaching power. As a docent Donner lectured on the history of English literature (1898, 1903-1904), German romanticism (1899), Lord Byron and his English contemporaries (1899, 1901), Goethe and Schiller (1902, 1903), the two of them and Lessing (1910-1911), the history of the novel from the sixteenth to the eighteenth centuries (1905), and Elizabethan drama (1906, 1907-1908).

In 1912 Donner moved to Sweden, where he soon took out Swedish citizenship. He continued to have an impact on Finnish academic life, however: through a large endowment he made it possible to create a professorship in English language and literature at the newly founded Åbo Akademi.

Hjalmar (or Jalmari) Hahl (1869-1929) came from an artist family, and showed considerable artistic talent in a number of fields: he conducted the Student Singers choir in the 1890s, for a short period after Kaarlo Bergbom's retirement in the early years of the century directed (with no small success) the National Theater, and published short stories and plays.[170] After teaching school in Helsinki for several years in the 1890s, Hahl went to graduate school at the Sorbonne, and defended his dissertation in 1896, a 209-page work entitled *Les tendances morales dans l'oeuvre de Giacomo Leopardi.* Hahl had originally planned to look at Leopardi's philosophical thought in its entirety, but there was such a wealth of secondary literature on Leopardi that he decided to narrow his focus to the moral tendencies in Leopardi's work. In the dissertation he argued that Leopardi's moral philosophy was not a systematic pessimism like Schopenhauer's or Hartmann's; rather his pessimism was a reflection of tragic events in his life. Söderhjelm, who served as official faculty opponent, claimed that Hahl had failed to notice that earlier scholarship, such as that by Paul Heyse, whom Hahl had neglected to mention, had already made this point.[171] Söderhjelm also found Hahl's arguments for an optimistic undercurrent in Leopardi's work unconvincing, and criticized the dissertation for its illogical argumentation in general. Hahl's discussion of more recent trends in moral philosophy was also superficial, Söderhjelm went on, and the dissertation was insufficiently grounded in a historical perspective. In this sense it resembled, Söderhjelm wrote, "a collection of scholarly vignettes about individual features of Leopardi's life and teaching."

In the first decade of the new century Hahl edited the sections on Icelandic, Danish, and Norwegian literature for the fourth volume of *Yleinen kirjallisuuden historia* ("General History of Literature," 1905); published, after his stint as director of the National Theater, a 600-page popular book on modern playwrights (*Nykyajan näytelmäkirjoittajia* ["Modern Playwrights"], 1908); and wrote articles for various publications in the field of aesthetics and modern literature. Partly on the basis of these qualifications, in 1909 he was appointed docent in aesthetics and modern literature, in which capacity he lectured at first (1910-1914) exclusively on Italian literature, then on French drama (1915-1916), aesthetic categories (1915-1916), and German drama (1918-1919). He gave his docent lectures alongside his job as a schoolteacher, and retired from the docentship two years before his death.

Gunnar Castrén (1878-1959) was raised in an upper-class Helsinki family, like Söderhjelm and Hirn in the progressive spirit of the eighties.[172] The young Castrén believed in European liberalism and a harmonious amity between the Svetophile and Fennophile camps; these ideals remained strong into his adulthood, when they were joined by a new belief in workers' rights. In the early decades of the new century, the period of Castrén's greatest scholarly activity, his Svetophilia became intensified, a shift that has been explained as the result of his bitterness at the deliberate exacerbation of language conflicts by the Fennophones (Enckell 6).

In his scholarly writings Castrén placed greater emphasis than either Söderhjelm or Hirn on contacts between native cultural Sveticism and Scandinavia. Castrén's powerful sense of tradition may have been due to his family: his grandfather was M. A. Castrén, and his great-aunt Sofi Kellgren, a contemporary of Runeberg's, introduced the young Castrén to the intellectual context of the mid-nineteenth century. "Nor did Runeberg alone come to life in these conversations, but my grandfather as well, and Snellman and Cygnaeus and Lönnrot," he later recalled (Enckell 4). Sofi Kellgren financed Castrén's doctoral studies at the turn of the century, and Castrén was later to write one of his finest cultural-historical studies of her deceased professor-husband, *Herman Kellgren* (1945).

Upon completion of his Master's degree in 1899 Castrén traveled south to visit European artistic and cultural centers, and in 1901 he lingered long in Stockholm researching the topic Estlander had suggested for his dissertation, *Frans Michael Franzén i [in] Finland*, which he defended in 1902. Living abroad reinforced Castrén's social radicalism; in Stockholm, for example, he became familiar with Branting's socialism and similar work. In the first decade of the twentieth century he played an active role in founding and writing for various Swedish-language cultural magazines, such as *Euterpe*, *Fria Ord*, and *Argus*; he was later to turn his cultural-political attentions toward other journals and publication series, including *Nya Argus*, the leading Swedish-language Finnish daily *Hufvudstadsbladet*, *Svenska Dagbladet*, and the Swedish Literature Society in Finland. Castrén frequently introduced magazine and newspaper readers to Finnish and foreign literature and literary criticism, and his articles contain numerous studies of individual critics and critical issues.[173]

Most of Castrén's own literary criticism focuses on either Finnish literature or congruences between Finnish and Swedish literature. This was even true in a sense of his 380-page dissertation on Franzén, in which he drew heavily on Franzén's collected poetry and new archival and epistolary sources in order to offer a detailed portrait of Franzén's life, personality, and work up to 1812, the year he emigrated from Finland to Sweden. Castrén's placement of Franzén's life and work in its cultural-historical and even social contexts is highly interesting. He notes Franzén's romantic nature and describes him as a preromantic poet: the desire for the peace of nature and hearth that shines through all his poetry mirrors the emergent German emotional inclinations (also centered around the home) that were building toward romanticism. The mentality was typical of the emergent bourgeoisie: the home becomes a focal point on which thoughts and actions are concentrated (348-349). Castrén also finds that special Finnish characteristic, melancholy, in Franzén's work, which thereby takes on a subtly elegiac quality that compares favorably with the colder, more satirical bent of Franzén's contemporary Mrs. Lenngren, who wrote on similar subjects. Hence Franzén's true significance lies in Finland, not Sweden. Castrén's thorough familiarity with Franzén's poetic mentality is evident for example in his reconstruction of the lost original of the "Sången över Creutz" poem, which won a Swedish Academy prize: his claim that the revisions required by Academy member Leopold destroyed the logical coherence of Franzén's original version was later confirmed when an early draft was discovered.[174]

It has been argued that Castrén was never a literary scholar with any great

Gunnar Castrén (M)

methodological or theoretical interest in his field.[175] This is partly true of the older Castrén, who never really understood the Swedish-language Finnish modernism of the 1920s (Enckell 13); but earlier in the century Castrén followed new methodological developments in literary criticism avidly. In his opening statement entitled "Litteraturpsykologisk forskning" ("Research in the Psychology of Literature") at his dissertation defense he offered a highly detailed account of the psychological study of literature.[176] He noted the shift in literary criticism from mid-nineteenth-century scientific models and the metaphysics and speculative aesthetics that had dominated the late nineteenth century toward a new psychological aesthetics whose central focus was on the process by which a work of art comes into being. This shift he saw as a reaction against nineteenth-century science: the unknown dimensions of the human psyche were now attracting scholars' attention. The current project was to determine as precisely as possible "the psychological undercurrents from which poetry springs," and to construct "using this and other available sources the writer's inward life, its structure and development, or even the shared mental states of entire periods." These objectives grew out of a conviction that literature is "the mirror image of the poet's inner life," and that a literary *oeuvre* constitutes for the psychologist of literature "a biography, parallel to that written in dates and facts," but more important, because the latter takes on significance only by virtue of the former. In studying a writer's inward development it was not as crucial to determine the origin of external impulses as it was to understand why the writer happened to be susceptible to those impulses.

This opening statement is something of a polemical piece in favor of the psychology of literature, which had been criticized for its inexact nature: the scholar must in the last analysis rely entirely on his imagination in order to create a living image of the writer's psyche, which placed the psychology of literature closer to art than to science. Castrén counters by arguing that this criterion would place all historicizing inquiry, even the study of history itself, beyond the bounds of science. Absolute objectivity and unimpeachably exact methods could only be expected of the natural sciences and those human sciences in which it was possible to use the methodology of the natural sciences, such as philology. In literary criticism, Castrén argued, the supposed neutrality or detachment of the scholar was self-delusion. Truly scientific inquiry is born not out of depersonalized passivity but rather through a constructive conceptualization that "discriminates and sifts and apprehends out of multiplicity that which is unified." This conceptualization in turn is restricted by the scholar's and his historical period's limits. "No literary-psychological work, therefore, or any other literary- or art-historical work, can ever be more than an expression of some individual's and to some extent also some period's understanding."

In addition to the dissertation and the late monograph on Herman Kellgren, Castrén's most significant scholarly works include *Stormaktstidens diktning* ("The Poetry of the Great Power Period," 1907), *Norden i den franska litteraturen* ("The Nordic Countries in French Literature," 1910), and the lengthy books on *Gustav Philip Creutz* (1917), *Juhani Aho* (1922, in two volumes), and *Runar Schildt* (1927). His only book on a foreign author is the short monograph *G. Bernard Shaw* (1906), in which he places particular emphasis on Shaw's social radicalism.

The Poetry of the Great Power Period is a stylistic history, based on archival research Castrén did at the university libraries of Helsinki and Uppsala and the Stockholm Royal Library, and secondary sources such as Hanselli's somewhat unreliable *Samlade vitterhetsarbeten af svenska författare från Stjernhjelm till Dalin* ("Collected Literary Works of Swedish Writers from Stjernhjelm to Dalin," 1856) and Evert Wrangel's and Arvid Hultin's studies of the same period, from 1888 and 1904 respectively. Castrén examines poetic motifs and stylistic features from Sweden's great power period, treating them as collective phenomena typical of the time. Indeed Castrén's selective criterion throughout is always what is most characteristic of the period rather than what is artistically most successful. After a discussion of the relation of Swedish literature of the great power period to world literature, Castrén deals thematically with erotic poetry, drinking songs, patriotic poetry, contemplative poetry, didactic and satirical poetry, and wedding and funeral verses, and offers an overall description of the poetic style and meters of the period.

According to Castrén the period in Sweden from the 1630s to the early eighteenth century was dominated, at the northern end of a general European trend, by the Renaissance poem – a poetic form stemming most clearly from Germany and Holland and stressing form almost at the expense of content.

In his *Norden i den franska litteraturen* ("The Nordic Countries in French Literature") Castrén first deals briefly with knowledge of the Nordic countries in ancient and medieval literature, then turns to portraits of Nordic culture in general and literature in particular in French travelogues and *belles lettres*. Central French portraits of the North Castrén finds in Mlle de La Force, P. H.

Mallet, Bernardin de Saint-Pierre, and Leconte de Lisle. But all in all Northern Europe had not played a significant role in French culture; if he had reversed the direction of his research concern, looked at images of France in Nordic literature, Castrén says, he would have had to examine "a large majority of Nordic cultural history," so heavily had this "cultural leader of the modern age" influenced the Nordic countries. Nevertheless Castrén describes his study as leading down to "a knowledge of literature's significant underground veins, which only weakly push up toward the surface, but which have not been without effect on the directions in which literature has developed." In any event Castrén claims that the French have found it difficult to understand the finer points of Northern nature or penetrate to the inner spheres of the Nordic mentality. Despite their positive interest and polite phrases, most French writers believed deep down that Northern Europeans are barbarians after all – just as the oldest medieval travelogues had claimed.

Castrén's most important book may be his 443-page biography of *Gustav Philip Creutz*, a poet who emigrated from Finland to Sweden and there made a brilliant career in politics and literature. The book is structured chronologically, beginning with Creutz's childhood and adolescence and concluding with Creutz's work in the Swedish court; but whereas Hultin had concentrated on the external events of Creutz's life without aesthetic analyses of his literary work, Castrén is interested throughout on the light analyses of the author's personality can shed on individual poems. Another typical feature of the book is its comparative approach. In his analysis of Creutz's best-known long poem, *Atis och Camilla*, Castrén emphasizes its connections with Racine, which Martin Lamm had noted earlier: the simple plot and the restriction of characterization virtually to two protagonists both show, he argues, the influence of Racine's tragedies. In a sense Castrén is close to Hultin here, who had found more authentic emotion in Creutz's major work than Swedish critics had; Oscar Wieselgren, for example, had criticized Hultin's book in a review for its overemphasis on the innocence and inwardness of the Arcadian eroticism Creutz portrays. The eighteenth century, Wieselgren wrote, simply wasn't a century of paradisiacal purity, and there was no need to read such purity into *Atis och Camilla* either. In his review of Castrén's book (collected in *Essayer och kritiker* in 1919), Fredrik Böök similarly criticized the vision of *Atis och Camilla* as a great Racinean passion play as simply unfounded, and argued that the poem's eroticism was more likely, like most eroticism of that century, playful sensuality. Indeed Henrik Schück found Castrén's suggestion of Voltaire's influence over Creutz more convincing than that of Racine: Diana's powerful, almost operatic concluding monologue can be seen as indebted to Voltaire's poem *La Henriade*.

It is difficult to determine whether the inclination to read Creutz's poem as either playful or serious reflects a difference in national character. In any event, the Creutz book clearly reflects Swedish-speaking Finnish culture of the 1910s in its insistence on the differences between Swedes and Swedish-speaking Finns. Underneath the pan-European veneer of Creutz's poetry Castrén claims to find characteristically Finnish traits: the avoidance of bombastic rhetoric, the passive observation of nature, and the inner harmony of the poetic spirit, all of which have always been more typical of Finnish than Swedish literature (432-433).

The book on Juhani Aho, which has also been translated into Finnish, was timed to qualify Castrén for the professorship Werner Söderhjelm vacated, which he and Viljo Tarkiainen both applied for. As early as his dissertation Castrén had declared as one of his aims discovering the creative person behind the literary work; here in the Aho biography he announces his intention simply "to find Aho's personality in his works."

Even in the early 1910s Castrén had been considered so well-qualified a scholar that his name had been mentioned as a possible alternative to Werner Söderhjelm for the new professorship in Finnish and comparative literature.[177] Castrén had deputized for Hirn in the academic year 1914-1915, for Söderhjelm in fall term 1918, and again for Hirn in spring term 1919. When Söderhjelm retired from the professorship in Finnish and comparative literature, the job title was changed to "domestic" or Finnish literature, and faced in 1924 with Castrén and Tarkiainen the division decided to divide the post into two and appoint both, with Castrén teaching the "domestic" literature written in Swedish and Tarkiainen teaching that written in Finnish. In 1938 Castrén's job title was changed to "svensk litteratur," Swedish literature.

Upon assuming the professorship in 1924 Castrén gave an inaugural address entitled "Från Taine till Gundolf" ("From Taine to Gundolf"), which was not published until Castrén's late article collection, *Humanister och humaniora* (1958). Stressing once again the literary scholar's subjectivity and right to his own experience of the text, he argued that Taine's explanations of literature had been grounded in the assumption that human psychology could be explained in terms of mechanical causal relations. The new approach, on the other hand, represented by Benedetto Croce and Friedrich Gundolf and influenced heavily by the philosophy of Henri Bergson, placed trust in intuition as the basis for the study of art.

Two of Castrén's greatest scholarly achievements in the professorial career that extended until 1945 were his supplement to Henrik Schück's history of Swedish literature and his critical edition of the works of Runeberg. In his 500-page seventh volume to Schück's *Illustrerad svensk litteraturhistoria* ("Illustrated History of Swedish Literature"), entitled "Den nya tiden" ("The Modern Period," 1932), dealing with Swedish literature from the 1870s to World War I, Castrén demonstrated his wide-ranging literary erudition and capacity for sketching large overviews. Castrén's annotations to Runeberg's lyrical poems alone run to almost 1000 pages and constitute one of his largest-scale scholarly projects. The first volume of his annotations appeared in 1938, the second in 1956, when Castrén was 78 years old.

Arvid Mörne (1876-1946) came from a middle-class family; after his graduation from high school in 1894 he studied literature at the university and devoted his free time to civic activities, especially those having to do with the Swedish-speaking peasantry and workers.[178] This sort of interest was relatively rare among turn-of-the-century Finnish literary critics. Aspelin, for example, wrote in his journal during a period of political unrest in the fall of 1905 that a delegation led by Mörne, supposedly representing the working class, had gone "with extreme arrogance" to demand the resignation of the senate (October 31, 1905; 58-60). Mörne was also an excellent poet whose poetry reflected both early-twentieth-century social radicalism and the anti-fascism of the 1930s – though the bulk of his work was more "poetic" than radical.

Mörne's dissertation *Josef Julius Wecksell* (1909), which he had been working on since the middle of the decade, was the first comprehensive study of this Swedish-speaking contemporary of Aleksis Kivi who, like his Finnish-speaking colleague, had died incurably insane. The 290-page dissertation was grounded in extensive documentation that Mörne had gathered from literary works, archives, interviews, and the index to newspaper articles that was in use for the first time at the University of Helsinki library. The first part of the dissertation is a "life and works" type of biographical study; the second and, from the perspective of literary history, more interesting part places Wecksell's major work, the historical drama *Daniel Hjort*, in its literary-historical contexts. Mörne's analysis is dominated by influence studies (indeed, Arvid Hultin had earlier attempted to determine the sources of *Daniel Hjort*), as he traces Wecksell's sources in historical and literary treatments of the Cudgel War (a peasant revolt in the late sixteenth century) and his literary models in Shakespeare and others, and attempts to show how *Daniel Hjort* influenced later Finnish writers. Mörne does many close comparative readings of texts and is able to demonstrate intertextual relationships that could even today prove fruitful for critical discussions of *Daniel Hjort*'s allusive connections in the literary scene of the 1860s. With a good deal of praise for the historical accuracy of Wecksell's play, Mörne notes that *Daniel Hjort* achieved a major breakthrough in Finnish portrayals of the Cudgel War: in the play we see the peasant revolt as the first freedom struggle of an oppressed group, the first flaring up of class conflict. This reading is methodologically somewhat problematic, since Mörne earlier criticizes nineteenth-century historical dramas for their anachronistic projections of contemporary ideas into past historical events, and here engages in very much the same kind of anachronistic projection of twentieth-century social radicalism into a nineteenth-century dramatization of a sixteenth-century event. Mörne returned to Wecksell in 1920, in *Nya Wecksell-studier*, publishing many of Wecksell's poetic and vaudeville texts and providing information on their origins and textual composition.

In his attempt to qualify himself for a docentship, Mörne chose a topic that, considering his Swedish-speaking background and the inflamed battles over the language question, seems rather surprising: Aleksis Kivi. His 167-page book *Alexis Kivi och hans roman "Seitsemän veljestä"* ("Aleksis Kivi and His Novel *Seven Brothers*," 1911), which appeared a year after Viljo Tarkiainen's dissertation on Kivi, provided the Swedish-speaking reader with a general overview of Kivi's life and work, but also offered new readings of the nature and significance of Kivi's *oeuvre.* Mörne expertly places Kivi's works in the context of other Nordic depictions of the peasantry, and evaluates Kivi's literary position in terms of Finnish literary ideals of the second half of the nineteenth century. The Runebergian poetic ideal and late romanticism virtually blinded contemporary readers to Kivi's true nature:

> In studying the literary physionomy of the 1860s in Finland and Sweden it is at first impossible to determine how Kivi could possibly have made such an unexpected leap out of aesthetic conventionality. It is clear that the large Finnish audience could not comprehend Kivi. Its standards of taste until the eighties were partly regulated by Runeberg's poetry, partly by late romanticism, whose pale autumnal blossoms

Arvid Mörne (M)

> spread all over the Swedish-speaking world, most thoroughly in Finland through the work of the gentle dreamer Topelius. (88)

In this book on Kivi Mörne's concern with social questions and democratic empowerment come once again to the fore. The heavily dialogical narrative structure of *Seven Brothers* makes it a "peasant novel," inasmuch as it utilizes techniques of oral narration typical of the common people (58), and Mörne explains the brevity and hardships of Kivi's career not individualistically, as earlier critics had (poverty, alcoholism, love problems), but in terms of Kivi's rootlessness in the society of his day (29). At the same time it is our good fortune, Mörne argues, that the ruling class and its culture remained alien to Kivi (150).

It seems clear that Mörne's social radicalism hurt his chances for a docentship at the university. He was not appointed to a docentship until his second application, after he had published a long article on the Finnish press between 1854 and 1856 (1913), the first of a series of articles that made him a pioneer of Finnish press history (Ulfvens 212-228). It was followed by a 91-page booklet entitled *Från "Saima" till "Vikingen"* ("From *Saima* to *Viking*," 1916), in which Mörne describes how the evolution of nationalism among the Swedish-speaking minority was reported in the Finnish press from the 1840s to the 1870s. Typical of Mörne's personal interests in the matter was the emphasis he places on press reports of Swedish-language public education.

As a docent in the history of Finnish literature Mörne lectured on the history of Finnish literature in Swedish (1913, 1915, 1916-1917), Wecksell (1914), and nineteenth-century Finnish literature (1915). There was probably more; the information given about Mörne's teaching in the printed teaching programs is incomplete.

Text-critical research was done on the literature of ancient Greece and Rome largely in the department of classical philology, but a classicist also emerged from the more traditional literary-critical discipline as well, Emil Zilliacus (1878-1961).[179] Zilliacus was appointed to a docentship in the history of literature at the University of Helsinki in 1909, and while he was later appointed professor extraordinaire of classical literature (1943-1945), in point of fact literary criticism only dominated his early career. In his later years he was above all a poet, an essayist, and an excellent Swedish translator of classical plays.

Zilliacus graduated from the Finnish-language classical lyceum in Viipuri in 1896, studied Greek, Latin, and aesthetics and modern literature at the university, where he took his Master's degree at the age of 21 in 1900. He seems to have been most strongly shaped at the university by Werner Söderhjelm, whom Zilliacus thanks in the preface to his 1905 dissertation, *Den nyare franska poesin och antiken* ("Recent French Poetry and Classical Antiquity"). The *Euterpe* circle also had a powerful shaping effect on his career. He was very interested in French literature, and in the *Euterpe* circle Jean Poirot, a university lecturer in French and an active member of the circle who also assiduously introduced Finnish readers to French letters in periodical articles, fanned that interest.[180] During the *Euterpe* years early in the new century Zilliacus's original plans to do postgraduate work in art history, for which he had already traveled to Italy in 1899, were changed in favor of literature; and in pursuit of his new plans Zilliacus spent a year in Paris beginning in the spring of 1903.

In his doctoral dissertation, which he mainly wrote in Paris in 1903 and 1904, Zilliacus concerns himself with classical motifs in the work of French authors from André Chénier at the end of the eighteenth century to the Symbolistes of the 1890s. In an elegant essayistic style he traces French writers' increasing awareness of classical antiquity, which culminates in the Parnasso group at mid-century – Leconte de Lisle and Hérédia. He argues that the attachment to classical antiquity met some profound need in the French character and thus catalyzed the creation of many artistically excellent works. One of the most significant aspects of the dissertation is Zilliacus's listing of the features that should dominate translations of classical texts – sparseness, concreteness, pregnancy of expression – which were later to direct his own translations. According to the official faculty opponent, Jean Poirot, there was a certain imbalance among the various sections of the dissertation – the available source material had led Zilliacus to devote a disproportionate amount of space to earlier writers, for example – but Zilliacus also demonstrated an excellent command of both French and classical literature.

Like the dissertation, Zilliacus's next significant works, *Giovanni Pascoli et l'antiquité* (1909) and *Die Sage von Gyges und Kandaules bei einigen modernen Dichtern* (1909), deal with the reception of classical antiquity among modern writers. In the former, which was reprinted three years later in Italy, edited and with commentary by A. Gandiglio and L. Vischi, Zilliacus looks at a contemporary Italian poet whose collection *Poemi Conviviali* (1904) was heavily steeped in the preclassical tradition of the old Greek epic and lyrical poetry. In the latter he traces the use of a popular story told by Herodotus, Plato, and Nikolaus of Damascus in a short story by Gautier and plays by Hebbel and Gide. Zilliacus is interested in what happens when a classical tale is modernized, and in what elements various writers emphasize and why. Gautier, for example, is most interested in the "color and formal beauty" of the ancient story, and has added certain psychological motivations to it.

In *Klassiska bilder* ("Classical Pictures," 1910) the Swedish scholar Harald Brising had unearthed new information on the development of ancient Greek sculpture, and in Finland Zilliacus soon became a similar pioneer in the field of ancient Greek literature. In *Grekisk lyrik* ("Greek Lyric Poetry," 1911) he discusses central Greek poets from Archilochus to Anacreon, and in later years he supplemented that work with *Grekiska epigram* ("Greek Epigrams," 1923), *Lans och lyra* ("The Lance and the Lyre," 1933), *Choros* (1939), and *Eros och Eris* ("Eros and Eris," 1948), all of which together constitute an essayistic history of the ancient Greek lyric.

Zilliacus's scholarly work, which could hardly compete in his own eyes with his excellent work as a poet, translator, and essayist, proved significant for Finnish literary criticism precisely because, in a time when the most vocal and public criticism was trumpeting nationalistic themes, it kept up scholarly connections with the European cultural tradition. Zilliacus was thus one of the scholars who carried on the work begun by Söderhjelm and Hirn.

The department was also reinforced in the aesthetics branch when Kaarle Sanfrid Laurila (1876-1947), who had defended his 251-page dissertation entitled *Versuch einer Stellungnahme zu den Hauptfragen der Kunstphilosophie* ("Essay on the Central Issues of the Philosophy of Art") in 1903, was appointed docent in the philosophy of art in 1910. After Hirn Laurila was the greatest proponent of emotionalistic aesthetics in Finland, and in his magnum opus, the two-volume, 840-page *Estetiikan peruskysymyksiä* ("Basic Questions in Aesthetics," 1918, 1919), he built his own aesthetic system on Volkelt's foundation. In his books and articles Laurila returned again and again to the socio-ethical concerns of literature. His central authority in this respect was Leo Tolstoy, whose influential "What is Art?" (translated into Finnish in 1898) had a wide-ranging impact on the Finnish intelligentsia. Like Tolstoy Laurila insists on the morality and ethicality of literature, and over the course of time Laurila's aesthetic theorizing was gradually transformed into ethicoaesthetic polemicizing (Huuhtanen 135-141). Indeed, in the 1920s and 1930s Laurila's views became more stringent across the board: as he increasingly joined forces with the theologians' conservative literary front, he began to write about the dangers of leftist politics and literary prurience. His last works are in fact most interesting as documents of the right's most aggressive attitudes toward new trends in literature and scholarship.

Other Finnish aestheticians in the early decades of this century included Fredrik Lindström (1873-1924), whose dissertation *Taiteen tarkoitus ja moraali* ("The Purpose and Morality of Art," 1914) is ideologically in line with Laurila, and Onni Okkonen (1886-1962), later professor of art history, whose *Taiteen alku* ("The Origin of Art," 1916) is closely tied to Hirn's work.

Emil Zilliacus (M)

Viljo Tarkiainen and the Renaissance of Finnish-Language Criticism

In the first decades of this century Swedish-speaking scholars still dominated the field of Finnish literary criticism. At the same time, however, a period of expansion was beginning within Finnish-language criticism which was to lead, after World War I, to numerous doctoral dissertations written in Finnish and about literature in Finnish. Like criticism in Swedish, criticism in Finnish was clearly driven by an attachment to the critic's own native language: Finnish-speaking critics studied Finnish-speaking writers, and the history of Finnish literature tended to be understood narrowly along the lines laid by Julius Krohn, as the history of literature in Finnish. The pioneer in this field in the early years of this century was Viljo Tarkiainen (1879-1951).[181]

Just as Söderhjelm and Hirn founded modern Finnish criticism in Swedish, so did Viljo Tarkiainen found modern Finnish criticism in Finnish. Tarkiainen's scholarly work was characterized by Fennophilia, a love of realism, and democratic tendencies; these were evident for example in his insistence on the significance of late-nineteenth-century folk writers, in particular his elevation of Aleksis Kivi to the very top of the Finnish literary pantheon. In his first work on literature, *Kansankirjailijoita katsomassa* ("Visiting Folk Writers," 1904, in Finnish), based largely on his own interviews with a number of self-taught Finnish folk writers, Tarkiainen polemically juxtaposes the earthy realism of simple folk writers with the "petit bourgeois platitudes" (Aino Malmberg), "otherworldly lifelessness and colorlessness" (Maila Talvio), or "saccharine sentimentality" (Immi Hellén) of certain turn-of-the-century society writers. He complains that this new "sophisticated" literature was beginning to displace the vital folk literature of the late nineteenth century, and continues:

> Given that the rubric "art writer" provides no guarantee of said writer's absolute artistic superiority, and that "folk writer" does not necessarily indicate a lower order of artistic achievement – though these are the general assumptions we bring to these two terms – the question arises: what in fact *is* an art writer and what is a folk writer, and what distinguishes the two? (80)[182]

According to Tarkiainen the most important difference between the art writer and the folk writer is the former's familiarity with the use of certain artistic devices. But from a realistic point of view this was not crucial. It was precisely the stylistic roughness of the folk writer's work that had produced honest, authentic literary close-ups of "what a Savo woman of the people is thinking," or how desperately people "fought for their very existence" in the crown forests of Iijoki, or how famine years had battered Ostrobothnian peasants into readiness "for modern civilization."

Tarkiainen was the son of a small farmer from the Savo region, who graduated from high school in the tiny inland city of Mikkeli in 1897 and entered the university to study Finnish, with minors in foreign languages and aesthetics and modern literature. He took his Master's degree in 1903, with a dual major in Finnish and modern literature. His Master's thesis, *Äänneopillinen tutkimus Juvan murteesta* ("A Phonetic Study of the Juva Dialect"), was so brilliant that it was published in book form in 1904. Tarkiainen had spent the end of 1899 in Paris supplementing his French studies, but in the end philology did not satisfy him, and on his second university travel grant he went to Copenhagen to familiarize himself with "the vital, modern, and original research being done by Danish literary historians, who pay particular attention to the general history of culture" (quoted in Kari Tarkiainen 72). We do not know whether Tarkiainen met Brandes on this trip, as he apparently planned, but he did attend Valdemar Vedel's lectures on Shakespeare.

Beginning in the first decade of the new century Tarkiainen actively translated literary texts into Finnish, wrote book and theater reviews for various Finnish-language periodicals, and played an active part in cultural politics, including the Students' Aesthetic Society, where he became in practice the last chairman in 1903 (Kari Tarkiainen 67-68).

Tarkiainen's earliest plan for research on Finnish literature had been to write a broad historical account of folk portraits in Finnish literature, as his teacher Aspelin had done with German literature. Soon after he finished his Master's degree, however, he rejected this idea and turned to Aleksis Kivi, on whose life and works he began to collect material both in the library and in Kivi's home region, Nurmijärvi. Tarkiainen began his scholarly career with his 164-page dissertation on Kivi's *Seven Brothers* (1910), and a significant part of his most central scholarly work throughout his career would continue to revolve around Kivi.

In many respects Kivi was a fruitful topic for Tarkiainen. He was a realist, he had risen from the ranks of the common people, his works dealt with the people he knew best, and he was eminently suitable as a symbolic representative of emergent Finnish-language culture. Kivi satisfied both Tarkiainen's Fennophilia and his democratic mentality. His 640-page biography of Kivi, *Aleksis Kivi* (1915), is a cornerstone of Kivi criticism and indeed a model for later biographical research. Tarkiainen also played a crucial role in the compilation of a critical edition of Kivi's *Collected Works* (1915-1919) and also in the publication of various commemorative works dealing with Kivi's life and works.

While Tarkiainen never explicitly delineated his methodological or theoretical principles, it is clear from both the dissertation and the Kivi biography that he was working in a Schererian mode. Far more than previous Finnish critics Tarkiainen stressed detailed textual analysis and the biographical events, experiences, models, and sources that shaped the composition of Kivi's works. A novel feature in Tarkiainen's work compared with earlier Finnish literary criticism was the precision with which he analyzed, for example, Kivi's motifs and their background.

The dissertation is characterized on the one hand by a genetic, Darwinistic approach that explores the work Kivi did leading up to *Seven Brothers*, the origins and composition of the novel, its relationship to reality and literary sources, and its reception and influence. On the other hand, Tarkiainen brings a formalistic, almost philological perspective to bear on the novel, as he analyzes its structure, characterization, nature descriptions, humor, and style. His stylistic analyses in particular are redolent of linguistics, as he lists long strings of sentences in illustration of his points.

There are two respects in which Tarkiainen slackens the positivistic rope of strict factual objectivity. One is in the realm of evaluation. Ranking *Seven Brothers* with the works of Cervantes and Gogol, he polemically juxtaposes Kivi with the "antiquity" of Runeberg as the great poet of an emergent Finnish culture, a literary giant behind whose novel "stands an entire people with their innermost instincts and aspirations" (163). The other is a certain conflict between objective observation and subjective response. Tarkiainen needs an integrating force to bind together his overabundance of factual material, and he finds this force in Kivi's artistic vision, which is accessible only through identificatory intuition. Tarkiainen had obviously discovered that positivistic data-gathering did not provide a sufficient foundation for a complex scholarly construction of reality. Significantly, in his review of the Finnish translation of Emil Nervander's biography of Fredrik Cygnaeus (1907), Tarkiainen criticizes Nervander for publishing a disorganized mass of empirical data without delving into Cygnaeus's "innermost being" and finding there something that might come to life in the modern reader's imagination. Tarkiainen argues that Cygnaeus's literary work should have served as the key to understanding his life (502).

The contemporary reception of Tarkiainen's dissertation reveals its polemical thrust and, more generally, the current battlefronts of Finnish literary life. In his 1911 review of the dissertation in *Aika*, Tarkiainen's faculty opponent Eliel Aspelin criticized him for perpetuating that "dogma of the youngest scholars," that they were the true discoverers of Kivi. Aspelin reminded

Viljo Tarkiainen (HKM)

readers that Kivi's plays had been immensely popular in the theater continuously since the 1870s. He also disapproved of Tarkiainen's juxtaposition of Kivi and Runeberg, especially the notion that Kivi portrayed the Finnish people authentically and honestly and Runeberg did so only through the distorting lens of romantic idealism.

The gist of the argument revolved around the language question: i.e., around whether the "national poet" of Finland could have written in Swedish. Aspelin doubts whether Kivi has enough "ideological substance" to displace Runeberg – "never mind genius." In order to take Runeberg's place as the national poet, Aspelin argues, Kivi should have been a better educated writer. Yrjö Hirn too reviewed Tarkiainen's dissertation in *Nya Argus*, four years later in 1915, and despite the polite and even positive tenor of the review it is clear that he did not approve of Tarkiainen's juxtaposition of Kivi and Runeberg.

Tarkiainen's response to Aspelin in the third issue of the 1911 *Aika* contains the germ of his later work on Kivi. Though he denies any desire to rob Runeberg of his national-poet "trademark" and give it to Kivi, the spirit of the article is precisely the reverse. Runeberg may well have the "great 'national ideals'" required by Aspelin, he says; still, Kivi is just as important to Finns, and above all he is closer to them, because his works are grounded in the Finnish temperament and speech, a feature that "some might regard as the crucial prerequisite for the Finnish heir to that title 'national poet.'"

Fennomania men led by Eliel Aspelin cleansing the Maid of Finland of all traces of Sveticism. Cartoon from the Swedish-language humor magazine *Fyren* in 1908.

In his biography of Kivi Tarkiainen expanded the detailed research he had begun on *Seven Brothers* to cover Kivi's entire life and literary output. Indeed, so thorough was his research that later Kivi scholars have been hard put to add anything to the empirical data Tarkiainen gathered. His analysis of Kivi's specific connections with previous literary works, his models and influences, is equally thorough. The central area not covered by Tarkiainen, and thus left entirely up to later Kivi scholars, was the mid-nineteenth-century social, ideological, and scholarly context of Kivi's work.[183]

The Kivi biography had an immense impact not only on Finnish literary criticism but on cultural politics as well. Tarkiainen's Kivi was a nature poet, a solitary genius who, far from the literary debates, poor and despised (elements that had featured in Kivi criticism before, though partly incorrectly), gave literary life to all the wonderful qualities of the Finnish common people: toughness, good-naturedness, and humor. As a native and "common" speaker of Finnish himself Tarkiainen understands these qualities thoroughly and places particular emphasis on them in his biography. The tendentious nature of Tarkiainen's emphases makes a good deal of sense in the context of the times, when Finnish-speaking culture was just beginning to build idols for itself and rise to the social level of Swedish-speaking culture. The Kivi of Tarkiainen's biography must in some sense be seen as a cultural-political creation, an ideal that met a powerful social need (significantly, a third edition was run in the spring of 1916, the year after the book first appeared) and that was to exert an immense influence over the following decades of Finnish literary criticism and, more generally, Finnish culture.

Several of Tarkiainen's other scholarly works were indirectly tied into his work on Kivi. For example, his only study of a non-Finnish author, *Miguel de Cervantes Saavedra* (1918), was a biography arising out of Tarkiainen's university lectures on a writer whose major work Tarkiainen argued was the nearest comparison to Kivi's *Seven Brothers*.

Tarkiainen wrote a number of books on folk poetry. Two of these, *Aino ja Kalevalan muut naiset* ("Aino and the Other Women of the *Kalevala*," 1911) and *"Jos mun tuttuni tulisi"* (the name of a Finnish folk poem, "If My Friend Should Come," 1912), were attempts to expand Tarkiainen's field of expertise for a docentship – though the latter had been written earlier, for Kaarle Krohn's seminar on folk poetry. A third, *Ritvalan Helka* ("Whitsuntide at Ritvala," 1922), was written to improve his chances at a professorship. Throughout his career Tarkiainen clung tenaciously to his conviction that the unwritten folk poem was the first form of Finnish literature.

Aino and the Other Women of the Kalevala is a study of the *Kalevala*. While Tarkiainen notes that "in a psychological perspective huge gaps still loom" in the epic between the people's own poetry and Lönnrot's edited version of it, in practice he fails to take these gaps (which have become so important to later Finnish *Kalevala* scholarship) into consideration. He grounds his inquiry on the notion of an original coherent but lost Finnish epic:

> The most perspicuous prosodist must rake up those fragmentary pieces and guess at their original form with as great an effort as was expended by Lemminkäinen's mother in raking the pieces of her son's body from the black Tuonela River and sewing him back together. (8)

Even without paying sufficient attention to Lönnrot's editorial work as compiler of the *Kalevala*, however, Tarkiainen does make many significant textual observations and uses them to construct coherent "Finnish" *Kalevala* characters:

> The women of the *Kalevala* have their own Finnish tonality. They do not have the heroic stature of the Scandinavian Viking women, or the brutality of the Slavs, or the sensuous glow of Mediterranean women. Their emotional makeup is gentler, softer; their attitudes toward life calmer, less passionate. Even in the hottest moments they remain relatively calm, turning all their emotions into tears, which they pour forth in abundance, even to excess, as indeed do the male heroes of the *Kalevala*. Let us recall for example the floods of tears shed by Aino's mother, Kullervo's mother, or the bride leaving her childhood home. Otherwise they rarely show their feelings outwardly. In this respect the *Kalevala* is probably one of the poorest among world epic poems. (98-99)

Tarkiainen's early work, particularly *Fredrik Cygnaeus runoilijana* ("Cygnaeus as Poet," 1911), *Minna Canth* (1921), and *Gustaf von Numers* (1922), reflects a continuous attempt to elicit new aspects of the writing under scrutiny. Thus for example Cygnaeus's literary writing, which Nervander almost entirely neglected in his biography, becomes for Tarkiainen a reflection of Cygnaeus's feelings of isolation and loneliness and must be understood "psychologically as the most important and central part of all" of Cygnaeus's public activities (11).

At the end of 1911 Tarkiainen was considered for a docentship in Finnish literature, but at that time, as we have seen, his application was rejected, and it was not until the next year, upon his reapplication, backed with two new publications in the field of Finnish folk poetry, that his appointment was approved. Tarkiainen received the docentship in January of 1913.

According to university class schedules, during his docentship (through 1919) Tarkiainen lectured on relatively few topics: the development of Finnish literature (1913, 1915-1916), Finnish realism (1914, 1916-1917), Aleksis Kivi (1914), and Finnish literature in Swedish (1915, 1917-1918, 1918-1919). He soon received more demanding scholarly tasks, however, when he was appointed Söderhjelm's deputy in the professorship for the acacemic year 1916-1917, then again in autumn term 1917, for the entire calendar year 1919, and then continually from 1920 till he was appointed to the Finnish-speaking half of the split professorship.

During his docentship Tarkiainen also served as official faculty opponent in several doctoral dissertation defenses, particularly when the dissertations dealt with literature in Finnish. The first dissertation for which he served as opponent was the second written in Finland on a Finnish-speaking writer, Aukusti Simelius's *August Ahlqvist runoilijana, arvostelijana ja tyyliniekkana* ("August Ahlqvist as Poet, Critic, and Stylist," 1914). During the next decade, with the boom of new literary criticism in Finnish, Tarkiainen became the most important director of and opponent for dissertations on literature in Finnish. The dissertations for which he served as official faculty opponent included Ragnar Öller's *Ett kvarts sekel av vårt litterära liv 1828-1853* ("A Quarter

Century of Our Literary Life," 1920), and in Finnish, Ilmari Havu's *Pietari Päivärinta* (1921), V. K. Trast's *Julius Krohn runoilijana* ("Krohn as Poet," 1922), Aarne Anttila's *Johannes Linnankosken "Ikuinen taistelu"* (Linnankoski's 'Eternal Battle,'" 1922), and Eino Salokas's *Maallinen arkkirunoutemme Ruotsin vallan aikana* ("Our Secular Broadsheet Poetry During Swedish Rule," 1923).[184] These dissertations clearly push up the crop that Tarkiainen had sown, with his own research and his polemics in favor of Finnish-language culture, in cultural soil enriched by national independence. Or, as Ilmari Havu intones in the concluding pages of his dissertation:

> He [Päivärinta] has helped to make Finland Finnish, to raise the level of consciousness of the generation whose sons and grandsons have borne arms to create a free Finland and thus also borne her name into distant parts of the globe. . . . In his literary works he left the Finnish people their own portrait for all to view and hold in their hands, for as long as they exist as a people and as Finns. (222-223)

In 1924 Tarkiainen became the second native speaker of Finnish (after V. A. Koskenniemi) to be appointed to a professorship in literature. His tenure as professor lasted until after World War II, during which time he wrote the massive *Suomalaisen kirjallisuuden historia* ("History of Finnish Literature"), of which more below.

Other Scholars in the Early Twentieth Century

At the turn of the century a number of scholars did doctoral work in literary criticism who later devoted their working lives to other fields, and who perhaps never had very high ambitions in the literary-critical field. The highest degree offered in literary criticism seemed eminently suitable for both librarians and educators. The following table shows the number of dissertations written at the turn of the century in the field of aesthetics and modern literature by decade:

Dissertations	in Swedish	in Finnish	Others	Total
1890 - 1899	2	1	2	5
1900 - 1909	8	-	1	9
1910 - 1919	2	5	1	8
1920 - 1929	7	7	1	15

There were a number of reasons for the increase in the numbers of literature dissertations. The active work of the Swedish Literature Society in the areas of publishing and cultural politics (the language question) gave literary criticism in Swedish a shot in the arm; and at the same time the emergent Finnish-speaking elite sought to reinforce their cultural identity through the historical study of literature in Finnish. Also, the women's liberation movement made it possible for the first time for women to be trained as literary scholars. Input into the field came also from Söderhjelm in Germanic and Romance philology and other, younger scholars, who were studying abroad in ever greater numbers – in Berlin, Paris, or other leading European university cities.

Söderhjelm's influence was seen also in the tendency within aesthetics and modern literature for students to choose dissertation topics in the Romance languages (Lönnbohm 1895, Hahl 1896, Törnudd 1898, Zilliacus 1905, Erich 1912, Lehtonen 1913) – and, frequently, to thank Söderhjelm in their prefaces.

The first French scholar around the turn of the century was Kasimir Lönnbohm (1866-1919), who went on to make a name for himself as a journalist, reviewer, and poet under the pseudonym Kasimir Leino (brother of the great Finnish poet Eino Leino).[185] Born in Northern Finland to a minor government official and his wife, Lönnbohm entered the university in 1885 to study aesthetics and modern literature and Finnish language and literature. In his very first year at the university his name appeared in public in connection with the realistic movement in literature that was just making its breakthrough: in the fall of 1885 his article entitled "Realismi ja idealismi" ("Realism and Idealism") appeared in the *Hämeen Sanomat*, one of the first significant polemics in favor of Finnish literary realism. In it Lönnbohm seems to have been influenced by a talk given by the Danish scholar Herman Bang in the spring of the same year.

Lönnbohm took his Master's degree in the winter of 1888, and Aspelin, then a docent, encouraged him to continue his literary studies at the postgraduate level. Lönnbohm spent a year in Berlin in 1889, familiarizing himself with German literature, and in spring term, 1890, he pursued further studies in Paris. Upon going abroad his plan had been to write a dissertation on modern German literature, but Aspelin had urged him to choose a French topic because, as Aspelin said, "Finns lack the German inclination toward long-windedness, and incline instead toward Gallic businesslike brevity" (quoted in Kaukonen 35-36). Söderhjelm's well-known preference for French literature also carried a good deal of weight.

During the spring of 1890, then, Lönnbohm collected material in Paris for a dissertation on Prosper Mérimée. For various reasons, however – for example, it transpired that Augustin Filon was currently working on a big book on Mérimée, which appeared in 1894 – the dissertation was delayed. After working for some years writing for papers, translating texts into Finnish, and working in the theater, Lönnbohm submitted his 258-page dissertation entitled *Prosper Mérimée* in 1895, dealing with "his life and works from a literary-historical perspective." The dissertation is written in a fluent journalistic style spiked with rhetorical questions and recursive and anticipatory "narratorial" remarks like "as we are already aware," "which we will soon see," and so on. Lönnbohm passes over the literary works quickly, pausing only to take a longer look at their public reception. Aspelin, who served as official faculty opponent, criticized the dissertation for this emphasis on biography at the expense of literary-historical analysis, but also praised the biographical work as richer in detail than "any other study of Mérimée published to date."[186] In his voting statement for the division discussion of Lönnbohm's dissertation, J. J. Tikkanen recommended its approval but also suggested that, rather than making Mérimée come to life, Lönnbohm had drowned his life in minor details. The extensive biographical data and its chronological presentation did render the dissertation scholarly, Tikkanen admitted.[187] Edvard Westermarck, on the other hand, recommended that it be failed, arguing that it was not a "literary-historical study" at all, simply an unscholarly biography.[188] The dissertation was approved with a vote of 9-2, but the severe criticism leveled against it in division discussions diminished Lönnbohm's chances of getting an appointment in the department.

Lönnbohm left for his second extended journey abroad in late 1896, traveling to Vienna, Munich, and Berlin. His primary goal was to familiarize himself with the theater, but he seems to have been planning further scholarly research with a docentship in mind. Lönnbohm was unable to concentrate on scholarly research alone, however. Starting in 1899 he spent three years as director of the regional theater in Viipuri. Toward the end of the first decade of the new century he began to turn toward art history, publishing a book in 1908 on Aleksander Lauréus, a Finnish-born painter who moved to Sweden and eventually became court painter. Soon after this mental illness cut short his career.

Alvar Saladin Törnudd's (1866-1943) 215-page dissertation, *Etienne Pivert de Senancour* (1898) was the first biography ever written of the French author. Törnudd had taken his Master's degree in 1889, and while studying in Paris he had familiarized himself with Senancour's manuscripts; in 1897 he published an article in the second volume of the Memoires de la Société Néophilologique (a society founded in 1891), entitled "Quelques feuilles d'un manuscrit inédit de Senancour," introducing readers to the manuscript of Senancour's *Méditations.*

In his dissertation Törnudd admits that the preromantic Senancour was not of the first rank as a writer, but insists that he nevertheless had "his place among the finest and best in the history of human nature" (ii). His central biographical source is the unpublished manuscript of the biography Senancour's daughter wrote of him in 1850. In addition to his discussion of Senancour's life and works, Törnudd compiled an annotated bibliography of all his known published works (199-215). Unfortunately the list is not complete: it lacks the early unpublished manuscript *Aldomen.* In his discussion of individual works, Törnudd devotes particular attention to Senancour's epistolary novel *Obermann,* which he compares with other *Werther* imitations while also seeing in it a strong autobiographical component.

In his statement as official faculty opponent, Söderhjelm praised the dissertation's biographical section, but found it lacking in comparative analysis of the literary *oeuvre*: Törnudd should have examined the literary models for *Obermann* and Senancour's connections with the encyclopedists, Bernardin de Saint-Pierre, and Malebranche. Törnudd explored his subject too much from within, and failed to draw all the conclusions which his extensive material allowed.

Alexis von Kraemer (1871-1927) was one of the most active scholars of French literature around the turn of the century. Born in St. Petersburg the son of an admiral, von Kraemer studied in Paris for several periods of time in the 1890s before settling in Helsinki, where he played an active part in academic and cultural life, first writing for *Euterpe,* then working in the library, finally teaching French at the university. His 196-page dissertation, *Villiers de l'Isle-Adam* (1900), is a detailed but fairly shallow and unscholarly analysis of the late-nineteenth-century French symbolist's life and works. In addition to the most recent Villiers criticism, von Kraemer drew on interviews he conducted with Villiers's contemporaries, for instance Stéphane Mallarmé; but in the end he makes very little attempt to place Villiers in the history of French literature. Indeed in his statement as official faculty opponent Söderhjelm noted that von Kraemer should have expanded his scope beyond the quite competent biography to situate Villiers in the literary and social milieu of his day.[189] The dissertation's significance as a Villiers biography was in fact enhanced by the paucity of previous Villiers criticism; prior to von Kraemer's work only one monograph had been written on him, by R. du Pontavice de Heyssey in 1893. Also of interest is von Kraemer's double vision of Villiers – like A. H. Bergholm's dissertation on Almqvist it is steeped in Finnish cultural and ideological assumptions of the turn of the century. Von Kraemer places a high value on Villiers's symbolism, his "religion," his satire, his critique of the petit bourgeoisie, and his indifference to naturalistic depiction:

> From the very first moments of his public life as a writer Villiers de l'Isle-Adam was aware of the defects of materialism and artificial realism, and convinced that the positivistic world view was incapable of offering a satisfactory explanation of existence. (157)

It was no coincidence that von Kraemer became one of the central figures in the *Euterpe* group, and that, in his articles on the most recent French literature, he stressed its impressionism, its "new worlds of beauty," and its stylistic virtuosity, complaining that it was too little known in Northern Europe.[190]

Mikko Erich's (1888-1948) 266-page dissertation *Ugo Foscolo come uomo e come poeta lirico* ("Foscolo as Man and Lyric Poet," 1912) focused on the Italian romantic Ugo Foscolo. Based on research Erich did on three trips to Italy around 1910, the dissertation asserts the significance of biographical factors in Foscolo's lyric poetry and reflects a good deal of admiration for Foscolo as a man and as a poet, but does not delve very deeply into the actual poems. Due to the cursory nature of Jalmari Hahl's statement as official faculty opponent, the division requested its other members to submit written statements as well. Hirn found the dissertation wanting in comparative literary-historical methods – comparison with other writers, such as André Chénier, using classical motifs – but recommended it be approved.[191] Söderhjelm, on the other hand, wanted to fail the dissertation. He argued that it was impossible to tell from Erich's text whether it was intended for a Finnish or an Italian audience, and found Erich's analyses to be rudimentary in the extreme, as if he were writing a practical guide to teaching Foscolo instead of a scholarly thesis. Foscolo's style in particular, Söderhjelm noted, offers a fertile field for scholarly analysis.[192] Erich's study was the first Finnish dissertation ever written in Italian, and this hampered its processing in division meetings. Professor of theoretical philosophy Arvi Grotenfelt had his frustration recorded in the minutes that the division's hands were virtually tied because so few of its members had any Italian. In the end the division approved the dissertation with a vote of 15-3.[193]

Erich's lengthy work *William Shakespeare ja hänen runoutensa* ("Shakespeare and His Poetry," 1916) was a popular Finnish introduction to Shakespeare's work grounded solidly in Shakespeare criticism. The book was timely, for it appeared shortly after Paavo Cajander had completed his Finnish translation of Shakespeare's plays. Erich was primarily an educator, a politician, and a lawyer, but even in the 1930s he published short little studies of Aleksis Kivi in the Society for Literary Scholars annuals.

One of the brightest names among the rising generation of Finnish-speaking critics was Johannes Vihtori Lehtonen (1883-1948), a farmer's son who completed his university degree in a record three years and then traveled to Germany and France for postgraduate studies.[194] Lehtonen's dissertation, *Tutkimuksia Théophile Gautiern runoaiheista* ("Studies in Théophile Gautier's Poetic Motifs," 1913), in the preface to which he thanks Gustave Lanson of Paris for suggestions and help, is a good example of the type of dissertation that became common in the early years of the century: it comprised a series of tightly interconnected essays or "studies" (tutkimuksia, studier). Lehtonen intended his 285-page dissertation to be mere the first series of such studies of Gautier, but it was also the last; no further volumes ever appeared. He is concerned in the work with "personal feeling," "love and woman," and the influence of folk poetry on Gautier's work. Lehtonen is particularly interested not only in Gautier's sources but also in his "inspirational connections," i.e., the powerful personal experiences that underlie his poems. Exposing this "deep-seated emotional response" (58), this "innermost fluctuations of feeling" (59), to the reader from behind Gautier's carefully constructed "pittoresque and plastic" form is one of Lehtonen's central aims in the dissertation. The dissertation is largely an explanation of biographical material and contains very little self-reflexive methodological commentary, but what little there is points to the same concern with the writer's personality:

> The task of historical literary research is to record the basic features of the poet's character, to observe his talent and its limits, and then to show how he expresses himself. More than this it cannot do if it would avoid subjective evaluations and their subjective competence. (221)

Lehtonen did later publish a few articles on Gautier, including one in a Festschrift for Werner Söderhjelm entitled "Théophile Gautier och antiken" ("Théophile Gautier and Classical Antiquity," 1919).

Lehtonen was a polyglot scholar whose familiarity with foreign literatures and critical traditions is clearly evidenced in the number and scope of his Finnish translations: between 1908 and 1924 he translated literary, essayistic, and scholarly texts from Swedish (e.g., Werner von Heidenstam, Yrjö Hirn), English (Walter Raleigh's book on Shakespeare, Locke, Ruskin, Scott), French (Victor Hugo, Montesquieu, Rousseau, Rochefoucauld, Taine), German (Gottfried Keller), and Spanish (Juan Valera) – a total of seventeen works.

After a popular biography of Victor Hugo in 1917 and a 58-page "Survey of the Main Trends in Nineteenth-Century European Literature" for the sixth volume of the Finnish *Maailmanhistoria* ("World History") in 1921, Lehtonen shifted exclusively to the study of Finnish literature. In the 1920s he became, alongside Tarkiainen and Rafael Koskimies (1898-1977), one of the leading Finnish-speaking literary scholars in the country, specializing, like Tarkiainen, in Aleksis Kivi. In his study *Runon kartanossa* ("In the Mansion of Poetry," 1928), for example, he took a close look at Kivi's poems, examining both their connections with themes and motifs from Finnish and world literature and their form and imagery. One significant byproduct of his research on Kivi was the collection he edited, *Aleksis Kivi aikalaistensa arvostelemana* ("Aleksis

Kivi as Received by His Contemporaries," 1931). Lehtonen was appointed docent of literary history at the University of Helsinki in 1923 and professor extraordinaire in 1935.

The wave of Fennicism in the first decade of the new century led in 1906 to the mass translation of Swedish surnames into Finnish (Aspelin, for example, hyphenated his name with the literal translation Haapkylä), and further polarized the literary-critical camps over the language question, so that, on the one hand, Finnish-speaking critics tended to restrict their choice of topics to programmatically Finnish ones (Tarkiainen), and on the other, Swedish-speaking critics began to polemicize for the cultural tradition carried by Finnish literature in Swedish (Hultin). This schism was not yet visible in the study of Swedish literature, however. In the early years of the twentieth century Nordic literature was the single most significant foreign focus of Finnish literary research. According to scholarship bibliographies for the first quarter of this century, considerably more research was done on Nordic literature than on Romance literatures, which was, for reasons discussed above, very much to the fore; and twice as much research was done on it than on German literature, with English and American literature falling well behind the German. In addition to the dissertations on Swedish literature by Boldt, Hultin, A. H. Bergholm, and Jenny af Forselles, and the monograph in Swedish on Norwegian literature by the historian Petrus Nordmann (1858-1923), *Björnstjerne Björnson* (1915), Nordic literature was assiduously presented in articles by Gunnar Castrén, Emil Hasselblatt, V. A. Koskenniemi, Irene Leopold (1860-1940), Agnes Langenskjöld (1887-1965), Werner Söderhjelm, and a woman who was later to become one of the leading book reviewers of her day, Anna-Maria Tallgren (1886-1949).

A good example of the unproblematic coexistence of Finnish-language literature and Swedish literature in the work of a single critic is provided by the scholarly work of A. H. Bergholm (1876-1949). In 1899 he had published a brief biography of the early Finnish-speaking romantic poet Abraham Poppius, along with a selection of his poetry, in the periodical *Suomi*, emphasizing Poppius's fidelity to his native tongue and his significance as a pioneer of Finnish-language literature. In his 1902 dissertation on C. J. L. Almqvist, however, he could still blithely turn to Swedish literature, even taking issue with Almqvist's Swedish critics for misunderstanding him. It would not be very many years before this sort of critical swim across the Gulf of Bothnia would be perceived as unthinkable. Bergholm argued against both the moralistic popular rejection of Almqvist and the treatment he had received at the hands of Swedish literary historians, pointing out that Almqvist had been appreciated in Finland from the start: among his many ardent Finnish admirers had been the J. V. Snellman of the period prior to the "Det går an" polemic. The dissertation is a kind of Finnish rehabilitation of Almqvist; Bergholm suggests that Almqvist's poetic spirit is more congenial to Finns than to Swedes:

> it seems as if the introverted Finnish character may have been more receptive to Almqvist's profound meditations and imaginations, his peregrinations through the world of fantasy, than the Swedish – as if, indeed, there were in Finland a stronger foundation for an appreciative reception of his ideas than in Sweden. (20)

Johannes Vihtori Lehtonen (HYK)

In his reconstruction of Almqvist's world view Bergholm relies more heavily on the letters than on the literary works. He exhaustively analyzes Almqvist's conception of literature and insists upon the affirmation and harmony of Almqvist's world view. There is a certain radical-individualistic neoromantic tone, typical of the turn of the century in Finland, that underlies Bergholm's discussions of Almqvist's high morality – of how it was misunderstood by his petty, limited contemporaries whose conventional moralism blinded them to his early-church Christianity, which "conjoins the worldly and the heavenly" (245). Almqvist's "spirit" had successfully opened before the eyes of ordinary people "the wonders of invisible worlds" (246).

Beginning in the 1880s the Finnish women's movement had been a significant force in social debate (see Jallinoja), and before long the effect of women's liberation began to be visible in scholarship as well. In the 1880s it was still possible for even women to say: "If a woman enters the academic factory to study at the university, she is subsequently neither woman nor man but a disagreeable mongrel form."[195] Despite much resistance, however, women did enter the university in this period, and early in the new century began to play a prominent role in Finnish literary criticism (see Rantonen). In the first years of the century three women defended their doctoral dissertations; and even without the doctoral degree Helena Westermarck, Lucina Hagman, and Helmi Krohn made significant contributions to Finnish literary criticism.

Helena Westermarck (1857-1938), better known as a writer and especially as a painter, took a feminist approach to literature, introducing Finnish and foreign women writers in articles in the *Finsk Tidskrift* and *Nutid*, and particularly in her monographs on George Eliot (1894) and Fredrika Runeberg, the writer-wife of J. L. Runeberg (1904). She also wrote a long introduction to the Finnish translation of Sara Wacklin's autobiography (1924). Her 1884 article in the *Finsk Tidskrift* on "Den realistiska riktningen i den franska romanen" ("The Realistic Tendency in the French Novel") made her one of the very first Finnish proponents of French realism.

Lucina Hagman (1853-1946), who also took a feminist approach to literature, was a teacher by trade, but also for many decades an influential activist in the women's movement and a member of Finland's first Parliament in 1907. In addition to her monograph on Fredrika Bremer, mentioned above, Hagman wrote a two-volume biography of Minna Canth (1906, 1911). It took her over ten years and is in many respects a pioneering achievement. It was the first lengthy (600-page) biography of a Finnish-speaking writer, and it is driven throughout by a clear feminist consciousness:

> Naturally, the worst count against her [Minna Canth] was that she was a woman. This meant that it was not enough merely to shout down *what* she spoke and wrote; it was also decried *that* she spoke and wrote. Due to her sex, in other words, she was doubly victimized. She was the "rotten worm" that must be spat out.
>
> It is understandable, therefore, that the attempt to depict Minna Canth's life may well give new life to old hatreds for her, and even redirect them toward the one who has undertaken this project. (1:vii)

Helmi Krohn (1871-1967) – internationally known as a children's writer under her pen name Helmi Setälä – wrote numerous popular biographies of early Finnish women writers and artists (Minna Canth, Isa Asp, Emilie Bergbom) and published a biography of Victoria Benedictsson entitled *Siipirikko* ("Broken Wing," 1915; 2nd ed. 1929). During her work on Benedictsson she corresponded at length with Georg Brandes.[196]

The first Finnish woman to complete a doctoral degree in literary criticism was Hanna Lindberg (1869-1909).[197] Lindberg graduated from the Helsinki Swedish Normal Lyceum in 1895 and completed her Master's degree at the University of Helsinki two years later. After her graduation she taught school for a while in Helsinki, translated some of Minna Canth's short stories and Teuvo Pakkala's novel *Elsa* into Swedish, and studied at the University of Berlin.[198] In her 1902 doctoral dissertation, *The Shrew. Argbiggans typ i den engelska literaturen intill Shakespere* ([!]"'The Shrew': The Figure of the Shrew in English Literature Through Shakespeare"), Lindberg traces images of women through medieval English literature up to Shakespeare, showing how male portrayals of women typically attempted to silence them: the active, vocal woman or "shrew" was portrayed as the embodiment of sin and contrasted with the idealized female type of the angel or madonna. The polemical tone of Lindberg's dissertation places it in the Finnish women's movement's first period of struggle, which is generally regarded as extending until the end of the first decade of the new century.

Jenny af Forselles (1869-1938) wrote her 1903 dissertation on the life of A. N. Clewberg, a Gustavian Finnish-born poet and learned man who became assistant director of the Stockholm Royal Theaters and member of the newly founded Swedish Academy in the late eighteenth century. Her research is grounded solidly in archival materials and shed important new light on the life of the theater in Stockholm in the last two decades of the eighteenth century. Af Forselles was a critical and independent scholar whose source-critical remarks (e.g., 38, 40, 147, 148, 233) were incisive and whose intervention on points where earlier criticism had fallen into ungrounded conclusions was decisive. In this sense she stands above those of her male contemporaries whose scholarly careers, like hers, ended at the dissertation. In his review of the dissertation Söderhjelm criticized af Forselles's tendency to draw hasty conclusions and her carelessness with quotations and details, but also praises her "mature, independent, and intelligent" analyses and finds her section on the theater an invaluable supplement to Oscar Levertin's *Teater och drama under Gustaf III* ("Theater and Drama During the Reign of Gustaf III").

Af Forselles did not pursue a scholarly career. She edited the Swedish writings of Elias Lönnrot for the Swedish Literature Society (in two volumes, 1908, 1911), and published a few articles on literature for various periodicals, but then devoted herself to teaching and politics, including a term as Member of Parliament.

Of these first women scholars in the twentieth century, Ruth Hedvall (1886-1944) fashioned for herself the most brilliant scholarly career. Hedvall was a modern stylistician who based her 343-page dissertation *Runebergs poetiska stil* ("Runeberg's Poetic Style," 1915) methodologically on an exact characterization of style. During her research for the dissertation she was in close contact with various Swedish scholars – she mentions Esaias Tegnér, Josua Mjöberg, and Fredrik Böök in her preface – and seems to have based her method on Swedish models. She employs quantitative linguistic analysis, but does not restrict her study of style to surface structures: style reflects not only the material aspect of language, but historically variable topics and thought structures as well.

> Let me clarify this with an example: it should be possible to demonstrate statistically that the neoromantics use color-words more frequently than the pseudoclassicists. But in order to explain this we must also confront the problem of content. As might be expected, the phenomenon in question is linked to landscape. The romantics situated their texts in an entirely different landscape than the pseudoclassicists, preferring to depict interesting and colorful lightings that modify natural color values, while the landscape we find in eighteenth-century poetry – especially early in the century, before romanticism began to influence classical writing – is typically lit by even and colorless daylight. And indeed what interest is there in noting what every reader already knows for himself: that the meadows are green and the sky is blue? (1)

Hedvall narrows the scope of her study to exclude Runeberg's metrics, since that had become "a scholarly discipline unto itself," and focuses on poetic structure, depictions of nature, Runebergian situations, and human portraits (chapter one), Runeberg's imagery, rhythms, and motifs (chapter two), and diction (chapter three). In this third chapter she applies the quantitative vocabulary analysis Mjöberg had developed for his studies of Tegnér to Runeberg. In his statement as official faculty opponent Söderhjelm praised both Hedvall's new stylistics, which went beyond mere dry statistics, and her perceptive and logically argued reading of individual poems. He also admitted that Hedvall had convincingly demonstrated Runeberg's interest in Herder before 1830, which he himself had missed in his own earlier work on Runeberg.[199]

Hedvall marketed her statistical stylistic method outside the dissertation as well: her 1915 article "Modern stilforskning" ("Modern Stylistics"), for example, is suffused with her faith in the potential of exact stylistic analysis. At the same time, she still insists that exact analysis be tied to the study of the writer's personality and world view: "Through the use of stylistic analysis we attempt to delve down into the central ideas of the poet's thought, the dominant moods of his emotional life."

Particularly in the 1910s, Hedvall published numerous literary-critical articles on Runeberg, some of them stylistic analyses, others influence studies. She also wrote a history of Swedish-language Finnish literature, *Finlands svenska litteratur* (1917), of which more later. In 1921 she published a brief monograph on Friedrich Hölderlin. Alongside her work as a schoolteacher, she continued her research on Runeberg right up until her death in the bombings of Helsinki in 1944, publishing her popular monograph *Johan Ludvig Runeberg och hans diktning* ("Runeberg and His Poetry") in 1931 (new edition in 1941).

In the 1910s two dissertations were defended on Russian literature, one in the department of Russian language and literature, the other in education. In his 355-page dissertation *Die Landschaft bei I. S. Turgenev* (1915), Hugo Tauno Salonen (1882-1940) attempts to distinguish between Turgenev's portrayals of landscape and nature description. Salonen argues that Turgenev's early work shows the influence of Byron and Goethe, while his later work shows the influence of Scott and Flaubert. In their discussion of the dissertation in a division meeting the literary critics among the members seem to have doubted Salonen's talents in the field of literary criticism: Söderhjelm and Hirn's deputy Castrén criticized him for his lack of artistic sensitivity, without which no literary study was possible. The education dissertation, *Leo Tolstoj och skolan* ("Tolstoy and the School," 1915) by Knut Arvid Laurent (1876-1965), reflected the Tolstoy boom of the early twentieth century, examining in 161 pages Tolstoy's educational views with particular attention to the theoretical stances he took in the *Yasnaya Polyana* journal in 1862 and various other forums.

University lecturers in Russian at the turn of the century also wrote and published numerous articles on Russian literature. Andrei Igelström (1860-1927), who was appointed to a lectureship in 1899, wrote *Valvoja* articles on Tolstoy (1898), Pushkin (1899), and Gogol (1902); Igelström's successor Vladimir Smirnov (1876-1952) introduces Leonid Andreyev to readers of *Päivä* in 1908 and other modern Russian literature in *Valvoja* articles in 1908, 1909, and 1914.

In the early years of the century women became an important force in Finnish literary criticism. The first female literary Ph.D. in Finland Hanna Lindberg (left, M), Jenny af Forselles (center, M), and Ruth Hedvall (right, ÅAB).

The chair in Russian language and literature gradually accumulated a sizable library, which upon Finland's independence became the largest collection of Russian literature outside the Soviet Union (see Widnäs).

By the 1910s Finnish-language literary criticism was finally beginning to establish a foothold for itself, due largely to the enthusiasm engendered for the Finnish language on the language-question front. Whereas in the first decade of this century not a single dissertation was written in Finnish, in the 1910s the number of dissertations in Finnish exceeded that of dissertations in Swedish, and in the 1920s and after the Finnish language continued to gain ground. With the creation of new linguistics professorships in the two languages and the increasing polarization (particularly at the University of Helsinki) of the language camps, literary criticism in Finnish and Swedish began to diverge sharply, and knowledge of literature across language barriers began to decline. It would not be until after World War II that a new, more congenial atmosphere would develop in which it would become possible to see Finnish literature and literary criticism as a single discipline, regardless of the language the literary and critical texts were written in.

Tarkiainen's cultural-political goals can be clearly seen in Aukusti Simelius's (1882-1959) 353-page dissertation, *August Ahlqvist as Poet, Reviewer, and Stylist* (1914, mentioned above). In the first volume of the projected three-volume series Simelius only deals with Ahlqvist's poetry, in fact; the second and third volumes never appeared. His position on the language question is clear: in his preface he asserts that "Finnish literature" had to date been but little studied, from Julius Krohn's literary history to Tarkiainen's work on Kivi. In other words, Simelius implicitly takes "Finnish literature" to include only literature written in Finnish. And indeed he examines Ahlqvist's poetry in terms of the development of Finnish-language literature: "He cleared the way for our serious poets and made it easier for later generations to follow" (340). Methodologically Simelius favors causal relations; he covers Ahlqvist's life in an introductory chapter intended for all three projected volumes, and then discusses his poems and their models and sources, his hymn lyrics, his Finnish translations of poems, his theoretical pronouncements on poetry, and his influence on later poets.

This period beginning in the 1910s also sees an increased interest in nineteenth-century Finnish-language literature; in addition to Tarkiainen's books and Simelius's dissertation it is evident in Kaarlo Nieminen's (1887-1955) study of nineteenth-century Finnish poetic meters and prosodic research, *Suomalaisen taiderunouden kielestä viime vuosisadalla (1890-luvulle)* ("On the Language of Finnish Poetry in the Last Century (Up to the 1890s)," 1913). It was also found in several new literary histories, whose rather one-sided portraits of Finnish literature were largely responsible for the increasing vogue among the Finnish-speaking educated class of the notion that the nineteenth-century history of Finnish literature was the history of literature written in Finnish.

Exceptional among Finnish-speaking scholars for his interest in international topics was Eino Cederberg (1889-1963), whose dissertation *Tuhlaajapoika-aihe* ("The Prodigal Son Motif," 1917) first surveys the use of the prodigal son motif in world drama in the early modern era, then compares the use of the motif in plays written for schoolchildren by two Nordic writers, the Swede Samuel Brask and the Dane Jens Lønborg. The methodological modernity of Cederberg's dissertation is evident in his refusal to proceed work by work, with simple summaries of each, as earlier scholars (Holstein, Spengler) had done; instead he divided the motif up into aspects, which he discussed one by one. This allows him to offer more specific evidence of the Nordic plays' grounding in the European tradition. In this light the better of the two plays, Brask's *Filius Prodigus*, becomes a polemical piece on various current issues – a critique of Catholicism, for example.

Cederberg did not ultimately pursue a scholarly career. He taught secondary school for many years, worked in various capacities in the state cultural bureaucracy, and published numerous patriotic children's books under a pen name.

J. A. Hollo (1885-1967), who would later become a professor of education and an important translator of world literary classics, wrote influential book reviews for a number of papers in the 1910s and early 1920s. He was originally planning to write his dissertation on Aleksis Kivi, but gave that up when he learned that Tarkiainen was already working on Kivi – even though the two articles he did eventually publish on Kivi in 1911 and 1912 suggest that his perspective would have been very different from Tarkiainen's. In "Huomioita Aleksis Kiven lyriikasta" ("Observations on Aleksis Kivi's Poetry"), he puts a certain ironic distance between himself and the main trends of early-twentieth-century literary criticism – i.e., the exploration of a work's basis in reality and of the "problematic depths" of the author's emotional life. Hollo is more inclined to study the "text itself" and the images it generates; he has been regarded as a precursor of the Finnish New Critics, who began to import their ideas from the United States in the 1950s.[200]

Despite the boom in Finnish-language literary criticism in the 1910s, the first Finnish-speaking professor of literature to be appointed in Finland came from outside the ranks of academic criticism. When the Finnish-language University of Turku was founded to offset the recently established Swedish-language Åbo Akademi, the professorship in Finnish and comparative literature was offered in 1921, without an official job search, to Veikko Antero Koskenniemi (1885-1962). Koskenniemi, one of the most prominent young Finnish poets of the day, had not completed a doctorate, but he was not entirely without academic qualifications in the field of literary criticism. In the 1910s he had published numerous articles in periodicals and essay collections on Nordic, German, and French literature, a short popular monograph on Alfred de Musset (1918), and a work entitled *Roomalaisia runoilijoita* ("Roman Poets," 1919), inspired by Emil Zilliacus's book on Greek poetry. Koskenniemi's most noteworthy scholarly achievements came later, in books on Goethe (1932, 1944), Kivi (1934), and the contemporary Finnish writer Maila Talvio (1946).

In his 1915 review of Tarkiainen's biography of Kivi Koskenniemi had also indirectly drawn his own self-portrait as a scholar. He argued that Tarkiainen's book lacked a liberating overview that "combines imagination with research." His own scholarly prose is exaggeratedly poetic, and particularly in his book on Kivi he indulges in the ahistorical Gundolfian mythologization of his writer's transcendent genius more unabashedly than any other Finnish scholar. In his portrait of Kivi time vanishes as inessential: Kivi is not primarily a product of his period, but surges toward the future "in an unconscious longing for new beauty and a new philosophy of life." Kivi's life in the world flattens out, loses force, becomes art. Kivi imitates nothing; the only important connection is between his works and "poetic vision." Koskenniemi's conception of the artist can be called aristocratic/autonomic: at its center one finds the great writers, geniuses, who soar above the heads of the masses and the pushing and shoving of everyday life, subject only to their own laws, and whose work opens readers' eyes to worlds that are, however nondenominational, still deeply religious.

Literary Historiography

The characteristic tone of early-twentieth-century literary histories in Finnish was nationalistic. Part of this was simply the continuing influence of Snellman's national idealism, but part of it was also a new quasi-nationalist wave that would still inform the two most significant literary histories written after independence, Viljo Tarkiainen's *Suomalaisen kirjallisuuden historia* ("The History of Finnish Literature," 1934) and Rafael Koskimies's three-volume *Elävä kansalliskirjallisuus* ("Living National Literature," 1944-1949). There seems to have been a surge of new interest in literary historiography in the early years of the century, and we can speculate that it arose on the one hand out of a need to bolster separatist linguistic/cultural identities, on the other out of the sociopolitical conditions of the Years of Oppression: in a period during which literary writers took stands on a dangerous political situation with the discursive tools of the historical novel, literary critics could also find powerful new possibilities in literary history.[201]

The principles of literary historiography were discussed in numerous articles during this period. In his 1901 inaugural address, "Kirjallisuushistorian tehtävästä" ("On the Function of Literary History"), Eliel Aspelin had argued –

no doubt with the recent appointment process and the peer review statements fresh in his mind – that internationally the status of literary history was growing and that the literary historian must approach his subject "on a large scale, in quest of large goals," without getting lost in philological minutiae. Aspelin assigns literary history an important national and human task: in concert with philosophy and general history it could show, better than any other discipline, how ideas and philosophies of life travel from people to people on their way "toward a dimly perceived perfection."

For Aspelin literary-historical phenomena were fundamentally a matter of international communication. A clearer vision of literary history appears in Arvi Grotenfelt's (1863-1941) 1909 *Valvoja* article "Historiallinen yhtenäisyys kansalliskirjallisuudessa" ("Historical Unity in National Literature"), which seemed to speak directly to the social needs generated by the Years of Oppression. In his article Grotenfelt insists that the literary historian is duty-bound to elevate the lasting national features of the Finnish literature he discusses. Like national art, he writes, national science must work to uphold the unity of national life and to raise readers' consciousness about that unity. While Grotenfelt agrees that historical change injects healthy new elements into the national character, literary history and the history of art in general (and indeed *all* historiography) must preserve the present's ties with the past. If these ties are severed and "the spiritual life of the people" is cast adrift from its mooring in history, the entire existence of the people is in jeopardy. Grotenfelt's article is steeped in Snellman's ideology of the *Volksgeist*: the best new works of art do not overthrow the old national ideals but "instead purify them into ever higher, nobler forms."

One of the focal issues in the debates of the 1910s concerned the status of the two national languages in the writing of literary history. A. H. Bergholm's central concern in his 1910 article, "Sananen kirjallishistoriamme kirjoittamisesta" ("A Word About the Writing of Our Literary History"), was whether national literature included literature written in both Finnish and Swedish. Both must be included, Bergholm argued, because they reflect different aspects and developmental stages of the same Finnish culture: "Ethnic discrimination in our literary history is an abstract theory that is utterly lacking in historical foundation." Bergholm noted that Swedish-speaking Finnish writers had highlighted the spiritual life and culture of the Finnish people as powerfully as had Finnish-speaking writers; Runeberg, for example, was "the noblest incarnation of the Finnish spirit."

The most lucid proponent of this view, that literature in both Finnish and Swedish should be dealt with in the same literary histories, among early-twentieth-century literary historians was Werner Söderhjelm. In his 1918 review of Ruth Hedvall's book *Finlands svenska litteratur* ("The Swedish-Language Literature of Finland," 1917), which was intended for school and reference use, he wrote that just as it was impossible to depict Finnish cultural history by restricting oneself to the accomplishments of only one language group, so too was it impossible to offer a satisfactory picture of a country's literature if one focused entirely on works written in a single language. On the contrary, Söderhjelm wrote, the literary historian must inquire into the ways and degrees to which one literature had influenced the other, the degree to which they differed from and perhaps complemented each other.

> One cannot say that either is in any way "more national" than the other, more indigenous to patriotic literature, and it is a simplistic formulation when a brief textbook survey of literature in Finnish this literature is called the only national literature, or when a similar work is entitled *The Stages of Finnish* National *Literature.*[202]

But Finnish literary historiography stubbornly refused to toe the line drawn by Bergholm and Söderhjelm for the entire first half of this century. Literature written in Finnish and Swedish was discussed in different works, and the massive literary histories by Tarkiainen and Koskimies drummed into the consciousness of the large public that the history of Finnish *national* literature was the history of Finnish-*language* literature. It was not until Unto Kupiainen's short textbook *Suomen kirjallisuuden vaiheet* ("The Stages of Finnish Literature," 1958) and the work of Kai Laitinen – *Suomen kirjallisuus 1917-1967* ("Finnish Literature 1917-1967") in 1967 and *Suomen kirjallisuuden historia* ("The History of Finnish Literature") in 1981[203] – that both language groups were examined on an equal basis.

Whether focused on one language or two, the literary histories of the early twentieth centuries all reflected the notion that the literary historian performed an invaluable social and national task. Against this ideological backdrop it is all the more surprising that the first literary history to be written in this period, the brief *Suomalaisen kirjallisuuden historia* ("History of Finnish Literature," 1910) written by the poet Eino Leino (1878-1926), should have thrown over the reigning assumptions and set up a historical periodization based on stylistic shifts. Leino's literary history is closely tied to the essay collection he had published the year before entitled *Suomalaisia kirjailijoita* ("Finnish Writers," 1909). In his brief vignettes on Finnish writers from Elias Lönnrot to the early twentieth century he had already, in that earlier work, sketched out preliminary literary-historical periods, which he now, in his specifically historical study, sought to refine and extend to the full range of literature in Finnish.

In the literary histories of the nineteenth century the stylistic trends of that century had not yet taken on clarity. Polén, for example, had paid no attention to literary romanticism; for him the period following 1831 was "the dawning of Finnish nationality." Godenhjelm calls the period from 1809 to 1835 "the Renaissance of the national spirit." He does mention Arwidsson's membership in the romantic school, and notes that Kivi was a "thoroughgoing realist" who had portrayed the life of the people in "realistic fashion," but he still fails to name romanticism and realism as collective phenomena that shaped and organized Finnish literature. Nor do the terms romanticism and realism appear in the table of contents in Krohn's literary history. Krohn does not mention romanticism in the text, either; he does describe realism as a Nordic trend that had influenced certain Finnish writers, such as Canth and Aho.

Compared with these nineteenth-century literary histories, Leino's book is strikingly modern. He delineates three distinct stylistic periods: the romantic period from 1835 to 1870, the realistic period from 1870 to 1897, and the neoromantic period from 1897 to the present. Where previous literary historians had used extraliterary criteria to periodize Finnish literature, or else had simply invoked the *Volksgeist* as a collective concept with which to group or distinguish texts, Leino arranges Finnish works purely under these rubrics of literary style.

Leino also uses the term national romanticism for romanticism, and sees it extending from the publication of the *Kalevala* to the publication of *Seven Brothers*, but also continuing on past Kivi's novel in the form of post-romanticism, "the last swells of which wash into our own time." Leino finds a Christian world view integral to romanticism. Realism he finds ranging from *Seven Brothers* to Juhani Aho's *Panu* (1897); integral to that trend is a scientific world view. In turn-of-the-century neoromanticism Leino sees a partial return to emotionally charged romanticism, but with the proviso that the neoromantics can no longer believe in the harmonious romantic view of mankind: "by the *fin de siecle* the integrated psyche had become fragmented for all time."

Leino introduces his periodization modestly:

> I am well aware that, before any comprehensive history of Finnish literature can be written, we must have more or less comprehensive studies of the most noteworthy individual phenomena in our literature. Nevertheless I have here, as in *Finnish Writers*, ventured to stray outside the boundaries marked off by previous literary historians and construe the periods of our literature in different fashion. I am likewise aware that to validate these periods in a binding scholarly way would require not only a competence that I lack but a good deal more time and resources than I have had at my disposal. I hope, however, that my work may serve as some sort of stepping stone toward a proper understanding of Finnish literature. (3)

Soon after the publication of Leino's book, O. A. Kallio (1874-1940) brought out his comprehensive two-volume history of recent Finnish literature, *Uudempi suomalainen kirjallisuus* ("Modern Finnish Literature," 1911, 1912; 2nd ed. 1928). Oskar Albin Kallio came from a poor rural background and, like Tarkiainen and the poet and translator Otto Manninen, graduated from the Mikkeli Lyceum.[204] He studied history, aesthetics and modern literature, and Finnish at the university; upon his graduation wrote for *Valvoja* for a while; completed a doctorate in Finnish history; and then taught secondary school in Savonlinna and Tampere. He published several educational books on cultural history and translated world classics (including Voltaire, Machiavelli, and Byron) into Finnish, but his two major works were literary histories, the above-mentioned *Modern Finnish Literature* and *Yleisen kirjallisuuden historian pääpiirteet* ("Principles of the General History of Literature," 1905). The former, a nearly 600-page "literary-historical source book and manual," deals with Finnish literature from the early 1860s to the present; the first chapter covers the "founding period" of Finnish literature, that preceding the realism of the 1880s. In the second section he discusses realism and neoromanticism, which he calls the "transitional period." The work serves, as Kallio himself suggests in his preface, as a kind of sequel to Krohn's *Stages of Finnish Literature*. And while his literary analyses and historical overviews are rather thin, the work is still a valuable guide to numerous now-forgotten Finnish authors of the late nineteenth and early twentieth centuries.

Ruth Hedvall's 296-page book *Finlands svenska litteratur* (1917), published by the University Committee set up to explore the founding of a

Swedish-speaking university in Turku, covers the history of Finnish literature written in Swedish from folk poetry to the early twentieth century, with particular emphasis on Topelius, Runeberg, Tavaststjerna, and Franzén. Hedvall was able to draw on an impressive body of early-twentieth-century literary-historical research, especially books and articles by Hultin, Söderhjelm, and Castrén; and indeed throughout the first part of her book she largely recapitulates other people's work, only venturing to offer original appraisals of the more recent literature. By way of introduction to each of her periods, Hedvall takes a quick look at the European literature of the day, and stresses the close interaction between Finnish and other European literature.

Hedvall's book was not received particularly well. Hultin criticized it for its neglect of late-nineteenth-century and early-twentieth-century cultural Sveticism, and Söderhjelm, as we saw above, deplored the absence of Finnish-language literature and the scanty depiction of cultural-historical contexts, such as literary journals and book reviews.[205]

Far and away the best Finnish literary history written in the first half of this century was Tarkiainen's *Suomalaisen kirjallisuuden historia* ("History of Finnish Literature," 1934). His conception of Finnish literary history had been adumbrated several decades before, in the articles written between 1906 and 1920 and collected in *Piirteitä suomalaisesta kirjallisuudesta* ("Features of Finnish Literature," 1922), on writers from Agricola to Linnankoski. In his introduction to the 1934 history Tarkiainen emphatically rejects the notion that Finnish literature is young and poor: Lönnrot's *Kalevala*, he writes, is a depiction of the early stages of the "Finnish *Volksgeist*," and while under Swedish rule "the special stamp of the Finnish spirit" did not in fact emerge, Aleksis Kivi rises "on an authentic Finnish foundation" to the level of the great masters of world literature. In interaction with other literatures Finnish literature has gradually assumed its characteristic national form:

> That originality of poetic talent and unique national flavor that typifies both our ancient folk poetry and our most recent literary productions gives us every right to look confidently to our literary future in the expectation that, now that Finland has gained her political independence, she will continue to defend her spiritual independence while yet remaining in close contact with the rest of human civilization. For the literary evolution of the nations, especially of small nations, does not move forward through isolation or national complacency. Still less does it advance through blind admiration for and slavish imitation of whatever excellence other nations may have to offer. Of all the vital conditions of poetry and all art, fidelity to oneself is the first and foremost. (18)

Like his predecessors, Tarkiainen restricts his scope to literature written in Finnish, but he does make passing mention of such writers as Juslenius, Porthan, Franzén, Arwidsson, Runeberg, Topelius, and Snellman, who wrote in Latin or Swedish, because "they have played an important role in the historical development of the Finnish national spirit, and most of them have had a profound influence on Finnish-language literature" (5).

In Tarkiainen's view, Finnish literature has a thousand-year history. It has had its high and low moments, largely because "the cultural and developmen-

Prominent members of the Finnish Writers' Union in 1920. Chairman Viljo Tarkiainen (left) with Iivo Härkönen (center) and Otto Manninen (right). (M)

tal status of Fennicism has been varied, as has their ability to receive and incorporate influences from without." In Finnish folk poetry Tarkiainen finds evidence of a defiant "Finnish character," "Finnish fortitude," "typical Finnish" humor, or expressions of "grim conditions" to which Finns have had to resign themselves in the Arctic North. Tarkiainen stresses these national features in his discussion of literary classics as well: Runeberg, for example, was "the apotheosis of the heroic spirit of the Finnish people"; Kivi was "the first brilliant apparition of the original poetic talent long dormant in the people"; Eino Leino wrote his best poems "in the spirit of ancient Finnish poetry."

Tarkiainen places literary works in historical perspective almost entirely in the introductory chapters to each main section, where he discusses Finnish history in general and the evolution of the "national instinct" in particular. Occasionally he will also make reference to the spirit of the times or to a particular political or ideological climate in connection with individual works, or in the course of making comparisons with foreign literatures.

In the conclusion to his work Tarkiainen sums up the special character of Finnish literature. In this he places himself in the tradition initiated by Godenhjelm and Leino, but also takes his characterizations far beyond those of his predecessors. Tarkiainen argues that Finnish literature is fundamentally a Nordic literature, charged with the ideas and ideals of Scandinavian and generally Western European and Swedish-speaking Finnish cultures. He finds the strongest parallels with Finnish literature in the literary traditions of Norway and Estonia. Nevertheless, Finnish literature has certain purely Finnish characteristics.

One distinctive feature Tarkiainen identifies in Finnish literature is a certain slowness to assimilate foreign stylistic and ideological trends – the Estonians, for example, move with far greater alacrity in this area. Another is the organic presence of nature:

> Nature speaks there in both its external and its internal sense. In Finland the rustle of the forest penetrates even the peripheries of the modern city; and many of our modern poets have remained entranced by the solitude and secrecy, the ancient magic and mystery of the forest. In their souls they find the blood heritage of their ancestors, the melancholy of the *Kalevala* and the *Kanteletar*, the inexplicable impulse toward isolation, or a premonition of the occult forces of nature. The Finnish psyche has been inclining in this direction since ancient times, to judge from both lyrical folk poems and spells and charms. Traces of this inclination are still found in printed Finnish literature, especially in the emotional lyricism and internal ecstacy by which most of our greatest poets have continued to delve deeply into the primitive soul of the people, the magic and romance of nature. (325-326)

Also distinctive of Finnish literature for Tarkiainen is a certain solemnity, simplicity, and pessimism of expression. Patriotic poetry, for example, has never been puffed up with chauvinistic vainglory, but rather steeped in a melancholy love for the people and the land. Finnish love poetry "rings with sadness," and the depiction of individual ambitions "is crowned with tragic awareness, if not utter pessimism and resignation." The most characteristic Finnish style, Tarkiainen argues, is realism, centered around rural life, the social existence of small villages (church, the vicarage), sometimes the cities (the upper crust, the middle class, the workers) or Finnish history. Only rarely has Finnish literature "ventured outside of the country's borders, on historical, imaginative, or real journeys to foreign lands" (327). Finnish literature has also shown a marked lack of interest in psychological or intellectual analysis.

Tarkiainen's work is grounded both in solid and comprehensive research and in a sensitive and thorough knowledge of Finnish literature. His inclination toward depictions of a stable Finnish character is itself also grounded in, or at least parallel with, theoretical trends outside Finland. One can discern in the subtext of Tarkiainen's writing Scherer's notions of the literary-historical significance of an idealistically conceived folk psychology, "the spirit of a nation," and Dilthey's insistence on the autonomy, self-sufficiency, and prophetic creativity of intellect alongside political reality.[206] One might say that these notions reached Finland late, but in another sense German literary theory provided Tarkiainen, like Koskimies somewhat later, with an excellent tool with which to examine literature in accordance with the needs of a young state just groping for its cultural identity.

Prior to independence Finnish scholars did not make nearly the advances in the field of world literary history that they made in the history of Finnish literature. In the nineteenth century Finnish students of literature had to make do with foreign histories of world literature. At the end of the nineteenth century the most commonly used textbook in this area was Johannes Scherr's *Allgemeine Geschichte der Literatur*, which some Finnish scholars spoke of translating into Finnish in the 1880s, notably Aspelin in an 1886 article in *Uusi*

Suometar. Brandes's *Hovedstrømninger* was also frequently assigned students to read in the original. Gunnar Castrén later recalled that he and his classmates were introduced to "the realms of world literature" by Brandes's book (*Humanister*, 376). Estlander's attempt to write a history of world literature had come to a halt at the turn of the century with the first volume, in the preface to which, however, he did in fact sketch out his conception of world literature. His aim was clearly to write a textbook for university students: he planned to cover the leading nations that succeeded classical antiquity, Provence, France, Spain, Italy, England, and Germany, and to exclude the Scandinavian and Slavic countries, since "their literatures are the provenance of other disciplines in our university" (1).

The first history of world literature to appear in Finland was the four-volume *Yleisen kirjallisuuden historia* ("History of General Literature," 1899-1905), only part of which was originally written by Finns. It was loosely based on a Danish work, heavily edited and partially rewritten or expanded by Finnish scholars: professor Knut Tallqvist (1865-1949), for example, wrote the section on Oriental literatures for the first volume (3-64), professor of Slavic philology Jooseppi Julius Mikkola (1866-1946) wrote the section on Slavic literatures for the fourth volume (391-629), except for Hungarian literature (631-690), which was covered by a Hungarian scholar living in Finland, Ferenc Zinnyei (1875-1947). Jalmari Hahl also edited the Nordic literature entries in the fourth volume (1-389) on Finnur Jonsson, Frederik Winkel Horn, Ph. Schweitzer, P. Hansen, Henrik Jaeger, Valdemar Vedel, Julius Clausen, and several others. The work remained in the course requirements at the University of Helsinki until the 1930s, when it was replaced by Eino Railo's (1884-1948) six-volume *Yleisen kirjallisuuden historia* ("History of General Literature," 1933-1937).[207]

Kallio's *Yleisen kirjallisuuden historian pääpiirteet* ("Principles of the General History of Literature") was the first history of world literature written in Finland. It drew heavily on the *History of General Literature* recently translated and edited from the Danish, partly also on Hermann Hettner's literary history, but Kallio had no great scholarly ambitions for the book; his aim was merely to provide Finnish readers with an accessible Finnish introduction to world literature. In the interchange he subsequently had with Olaf Homén over the work's scholarly credentials, Kallio remained firm: he never intended the book for "hyperaestheticians and cultural eggheads locked up in the ivory tower of their aesthetic and literary-historical knowledge."

After Finland had gained its independence, Yrjö Hirn worked with the Otava Publishing House to commission and publish a Finnish history of world literature (Riikonen, 26-27). This project never came to fruition in its entirety, but in addition to Tarkiainen's *History of Finnish Literature* it did produce two other volumes in the 1930s, one a history of German and Dutch literature by (respectively) Rafael Koskimies and Henrik Hartwijk (1936), the other a history of Estonian and Hungarian literature by (respectively) Friedebert Tuglas and Aarni Penttilä.

Notes

1. The two-volume Cygnaeus biography is by Rafael Forsman from 1923-1925; there are two Ahlqvist biographies, by A. J. Sarlin in 1924 and by Ilmari Kohtamäki in 1956; the Estlander biography is by M. G. Schybergson from 1916; the two-volume Hirn biography is by Irma Rantavaara from 1977-1979; the Söderhjelm by Henning Söderhjelm from 1960; and the Tarkiainen by Kari Tarkiainen from 1987.
2. Unless otherwise indicated, biographical information about the literary critics discussed in this book have been obtained from Carpelan, Carpelan-Tudeer, the *Biografinen nimikirja* ("Biographical Register"), and the *Kansallinen elämäkerrasto* ("National Biographical Collection").
3. The following table is adapted from Tiitinen.
4. This figure does not include August Ahlqvist's six literary-critical studies.
5. For discussions of the literary-critical research done in this early period, see the work done by Heikel and Hultin.
6. For a discussion of Svenonius, see Salminen.
7. For this see Arvid Hultin's 1914 study of literary and learning history in Finland, 27-28.
8. For a discussion of *Porthan and Classical Scholarship*, see Kajanto.
9. See Jorma Vallinkoski's two-volume catalogue of the Academy dissertations from 1642 to 1828.
10. See Hultin's 1918 study of literary studies under Porthan.
11. Sarajas's dissertation is also available in Swedish translation.
12. See Hautala's research on Finnish folklore research, in Finnish in 1954 (88-108) and in English in 1968 (19-26).
13. See Ilpo Tiitinen's 1986 article on economics and poetics, 67-70.
14. See Vakkari's "Roots of Library Science in Historia Litteraria" and Mustelin's account from 1969-70 of the transition from the history of learning to the history of literature.
15. See Hultin's long 1918 article on literary studies in Porthan's era, p. 5, and Tarkiainen's 1947 article on Weman.
16. See Häkli's essay in the 1983 fascimile reprint of Alopaeus's *Animadversiones.*
17. On these appointments, see Autio, 247-255.
18. For a discussion of Pipping's work in the library, see Nohrström, 48-105.
19. See Mustelin's 1971 article on the transition from the history of learning to the history of literature, 323-324.
20. See Hugo E. Pipping's article on Blomqvist.

21. For further discussion of Blomqvist's teaching, see Mustelin's 1971 article on the transition from the history of learning to the history of literature, and Hannes Sihvo's article on Blomqvist.
22. See Homén's 1918 article on Törnegren.
23. See Aarno Maliniemi's 1939 edition of Elmgren's notes, 369-371.
24. See Arvid Mörne's 1916 monograph on Swedish nationalism in the Finnish press, 33-38.
25. See Nohrström, 46-52.
26. For useful discussions of Hegelianism in Finland, see Juha Manninen's articles of 1979 and 1983.
27. See 67-90 in the 1817 issue, 93-129 in the 1818 issue.
28. See Bennich-Björkman 1970.
29. See Elmgren's 1848 article.
30. See Penttilä's article on the 1836 hymns and Renvall.
31. For discussion, see Karkama 95, 109.
32. See Werner Söderhjelm's discussion of this influence in volume 9 of his 1925 collection of his writings, 137-163.
33. Cf. Nilsson, Wennås, and Emin Tengström.
34. On the great Swedish-language poet Runeberg as classicist, see Tideström 60-70.
35. See volume 9 of Söderhjelm's writings, 23, and Gunnar Castrén's 1940 *Hufvudstadsbladet* article on the University of Helsinki and Scandinavian culture.
36. See Henrik Grönroos's 1944 article on eighteenth-century Finns in relation to Germans and Russians.
37. See Esko Häkli's afterword to the 1981 reprint of Magnus Björkenheim's work on old French literature in Finland.
38. See Sjögren's academic memoirs, 19, and Juha Manninen's monograph on J. V. Snellman, 31.
39. Annamari Sarajas traces the interest in English literature in Finland in a useful 1957 newspaper article.
40. See M. G. Schybergson's 1891 monograph on the history of learning at the University of Turku, 153-156.
41. For discussion, see Gunnar Suolahti's monograph on Yrjö Koskinen, 137.
42. See Suolahti's book on Koskinen, 136-141, and Klinge.
43. For the creation of the Uppsala post, see Gunnar Castrén's 1936 article on Runeberg and the university and Lars Gustafsson's 1986 monograph on aesthetics and literary history in Uppsala.
44. For debate on this issue in the senate from 1836 to 1839, see the chancellor's office records for 1839-1840.
45. See Castrén, "Runeberg," 142.
46. See Pentti Aalto on classical studies in Finland, 32-33, Veli-Matti Autio, 179-180, and Castrén on Runeberg, 133-148.
47. In his history of the history of learning in Turku and Helsinki, Olof Mustelin suggests that Nervander may have placed particular emphasis on literary studies (317); see also Autio, 252-253, and Emil Nervander's biography of Cygnaeus, 264.
48. See his "Små häften" of 1868, 11-17.

49. The senate discussed Laurell's proposal on August 16, 1848, January 15, 1849, and April 14, 1849.
50. This was not a new idea; the impossibility of one person teaching the history of learning in all disciplines had been argued twenty years before, in 1828. See Autio, 249.
51. For Cygnaeus's life, see Elmgren's contemporary notes and the biographies by Rafael Forsman and Emil Nervander.
52. Minutes of the Division of History and Linguistics, October 24, 1853.
53. Minutes, and report to the chancellor, of the university senate, October 29, 1853.
54. See Erik Kihlman's 1923 article on Cygnaeus, 79.
55. This point is stressed by Eino Karhu in his volume on Finnish literature, 350-353.
56. For a biography of Berndtson, see Carita Björkstén's long article.
57. See the *Morgonbladet* report on his popular lectures, April 2, 1850.
58. Berndtson's letter to the university senate, December 10, 1852.
59. For a biography of Clasen, see Ragnar Öller's article.
60. For a discussion of Toppelius's work in the theater, see Eliel Aspelin-Haapkylä's history of the Finnish Theater (1:25-32) and Yrjö Hirn's book on the theater in nineteenth-century Finland (263-280).
61. For Calamnius's biography see Colliander 644, and Kauppinen 193-195.
62. See Ahlqvist's 1872 review of the translation in *Kieletär*, 47.
63. As Eero Saarenheimo points out, the student group that called itself the Arcadia Association had attempted to develop an aesthetic terminology in Finnish back in the 1860s, but without much luck (13-14).
64. For Elmgren's biography, see Aarno Maliniemi's foreword to his 1939 edition of Elmgren's notes, v-xv, and Palmén's lengthy 1922 article on him.
65. For a discussion of Elmgren's work in the library, see Nohrström 107-125.
66. See Elmgren's notes, 550, and Jorma Vallinkoski's 1949 article on the history of the Finnish dissertation, 10.
67. For a discussion of Elmgren's work on Porthan, see Matinolli.
68. For Ingelius's biography, see Nordman and Elmgren's notes.
69. For Polén's biography, see Hytönen's early study of him and Laitinen's 1988 article on him.
70. Cited in Arppe, 28-29. There had been some controversy over the Society's negative attitude toward Finnish in the 1840s as well; see for example the *Morgonbladet* article "Finska språket och Finska Vetenskaps-Societen" on November 12, 1849, and Georg Schauman's 1916 article.
71. See the *Suometar* article for May 14 and 21, 1858, entitled "Yliopistollinen Taistelu-tilaisuus," "The Academic Fray."
72. Polén bolsters his self-defense subliminally here by alluding to the *Kalevala*. He has "Minä kynnän kyyhättelen," "I plow, I _____" (none of the Finnish lexicographical sources seems to know what "kyyhätellä" means; in Kirby's translation given below, the translator has assimilated it to "kyyhötellä," to stoop, huddle, crouch), an allusion to the passage in the second song of the *Kalevala*, where Väinämöinen is planting the bare soil of the new creation:

"Minä kylvän kyyhättelen
Luojan sormien lomitse, käen kautta kaikkivallan
tälle maalle kasvavalle, ahollen ylenevälle." (19)

"Now I stoop the seed to scatter,
As from the Creator's fingers,
    From the hand of Him Almighty;
That the country may be fertile,
    And the corn may grow and flourish."
        (Kirby's translation, 2:295-300)

The point of Polén's allusion is, of course, that, difficult or not, his pioneering work partakes of the divine creation *ex nihilo* that the romantics assigned to the poet.

73. For Krohn's biography, see Trast.
74. For Lönnrot's life, see Aarne Anttila's 1985 biography.
75. See Ahlqvist's 1884 study of the suggested hymnal revision, 28.
76. For Ahlqvist's life, see the biographies by Kohtamäki, Sarlin, and Simelius.
77. For Rancken's life, see Wikman's 1931 article on him.
78. For Lagus's life, see Homén's 1921 biography.
79. For Estlander's life, see his own 1918 memoirs, *Ungdomsminnen*, and the studies by Lindman, Mustelin, Schybergson, and Steinby.
80. See his memoirs, *Ungdomsminnen*, 232-234.
81. Schybergson tells the story in his 1916 biography of Estlander, 228.
82. Senate minutes, January 20, 1859.
83. See for example Pentti Aalto's book on modern language studies, 25.
84. See Estlander's 1905 article on Mistral.
85. For a discussion of Estlander as art historian, see Ringbom, 51-56.
86. See Aalto on modern language studies, 27-28.
87. Estlander's letter to the university senate, October 25, 1867.
88. Minutes of the Division of History and Linguistics, May 30, 1868.
89. For a discussion of Estlander's cultural-political ideas, see Lindman, 127-159.
90. See Schybergson's biography, 381-382.
91. See Barck's 1953 book, 21-22.
92. See Schybergson's biography, 289-296.
93. For Bergbom's biography, see Aspelin-Haapkylä's four-volume history of the Finnish theater, passim.
94. For a discussion of Heine's Finnish reception, see Irja Maliniemi.
95. For a discussion of Bergbom's views on the theater, see Verho.
96. See Cygnaeus's 1868 collection of essays on literature and art, 36-37.
97. See Aspelin-Haapkylä's history of the Finnish Theater, 1:235.
98. Quoted in Aspelin-Haapkylä's history of the Finnish Theater, 1:237.
99. Minutes of the Finnish Literature Society board, May 7, 1879.
100. Quoted in Aspelin-Haapkylä's history of the Finnish Theater, 3:17.
101. Quoted in Aspelin-Haapkylä's history of the Finnish Theater, 3:17.
102. See also Schmeidler.

103. See Jouko Hautala on Finnish folklore research, 67-76.
104. Quoted in Martti Haavio's 1949 article on the *Kalevala* cult, 240.
105. For a discussion of Snellman's writings on the *Kalevala*, see Annamari Sarajas's 1984 article.
106. See Pentti Karkama's recent book on Snellman's literary politics and Annamari Sarajas's 1984 article on Snellman and the *Kalevala*.
107. For a discussion of Grimm and Finland, see Kunze.
108. Quoted by Sarajas in her article on Snellman and the *Kalevala*, 40.
109. For discussions of Perander, see Laurila "Die Ästhetik" and Tiitinen's article on him from 1971.
110. *Suomalaisen kirjallisuuden historia. I. Kalevala.* This volume appeared in Swedish translation under the title *Finska litteraturens historia. 1. Kalevala* in 1891.
111. This section appeared in German translation under the title *Die Kalevala vom ästhetischen Standpunkt betrachet* in 1892 (Leipzig).
112. For a discussion of Vasenius's life, see Werner Söderhjelm's 1928 article on him; for a discussion of his work in the university library, see Nohrström, 126-134.
113. Note that later Frese critics, such as the Swede Evert Wrangel in 1895 and the Finn Arvid Hultin in 1906, have found much less evidence of Pietistic influence in his poetry.
114. See Estlander's 1880 review of the Ibsen dissertation, Bolin's 1883 review of the second Ibsen monograph, and Lönnbeck's 1884 review of the Frese book.
115. Cf. also Yrjö Hirn's 1935 monograph on the Runeberg cult.
116. For a discussion of the aesthetics problem in the job search, see Huuhtanen 97-103.
117. Minutes for October 12, 1900.
118. For a discussion of Aspelin as art historian, see Ringbom 56-61.
119. For Aspelin's life, see Lindström, 37-58, and Saarenheimo.
120. Ilpo Tiitinen makes this clear in his 1982 index of Finnish literary critics, 13.
121. See Koskenniemi's 1947 memoir of his student days, 63-64.
122. For discussion of Lönnbeck, see Kai Laitinen's 1980 article on literary history, 22-24.
123. See Tarkiainen's 1931 article on literary studies, 51.
124. See Aalto's volume on modern language studies, 150-151.
125. The paper was published in Neiglick's posthumous collection *Valda uppsatser* ("Selected Essays," 1916). For Neiglick's life, see Mustelin's 1966 biography.
126. See Aalto's volume on classical studies, 30-119.
127. See Aalto's volume on modern languages, 137-140, and Sarajas's 1968 study of Finnish and Russian realism, 27-34, 73-74.
128. For a discussion of Mandelstam's work, see Soini 83-95.
129. For discussions of Ernst Lagus's life, see Landtman's biography and Steinby, 200-214.
130. Cf. Hirn's 1935 monograph on this cult.
131. Cf. Hautala's 1968 volume on folklore research, 89.

132. For the debates on Estlander's proposals, see the university senate minutes for April 25 and November 10, 1877, and the division of history and linguistics minutes for May 9 and October 24, 1877.
133. See Gustafsson's 1983 monograph on Schück, 11-15.
134. Castrén's statement is collected in his 1958 book *Humanister och humaniora*, 97-104.
135. For the *Euterpe* circle see Mustelin's 1963 monograph; for the Students' Aesthetic Society see Huuhtanen, 104-107.
136. For Söderhjelm's life, see Castrén's *Humanister*, Hugo Suolahti's 1931 memorium paper, and Willner's 1971 biography.
137. See Henning Söderhjelm's 1960 memoir, 40.
138. Cited in Henning Söderhjelm's memoir, 56.
139. As reported by Castrén in *Humanister*, 17.
140. Reported by Henning Söderhjelm, 110. For a discussion of Söderhjelm's contacts with Paris in the context of philological research in the 1890s, see Aalto's volume on modern language studies, 31-42.
141. Cited by Henning Söderhjelm, 42.
142. As reported by Henning Söderhjelm, 42.
143. In the 1905 *Euterpe* article on recent Swedish studies of Runeberg, 340.
144. The book was translated into Dutch, with the title *De Italiaanische Renaissance*, in 1909.
145. By Tigerstedt in a 1939 article on Finnish literary science, 364.
146. Cited in Henning Söderhjelm, 193-194.
147. Rantavaara argues in her biography of Hirn that the reviewers were not unanimous at all, and that they submitted a joint statement because that was the only way Vedel could be made to back the Swedes' candidate Söderhjelm (1:249).
148. Swedish academic circles in particular noted the language battles underlying the job search, and, as Karl Warburg wrote in Gothenburg in May, 1910, a good deal of bitter irony was found in the fact that the professors who wanted to overturn the Swedish reviewers' statement were precisely those who sought to defend the status of Swedish in Finland. Hirn responded later that same month in two articles published in Sweden, one in the same paper in which Warburg's remarks had appeared, saying that he had given no cause for polarization on the language question.
149. Division minutes for October 15, 1910.
150. Division minutes, February 8, 1911.
151. Division minutes, December 9, 1911.
152. Senate minutes, January 31, 1912.
153. Division minutes, December 9, 1911.
154. The standard biography of Hirn is Rantavaara's two-volume work.
155. See Huuhtanen for a discussion of Hirn's aesthetic thought.
156. Tigerstedt in his 1939 article on Finnish literary criticism, 366.
157. See Tiitinen's 1977 article on psychoanalysis in Finland.
158. See Mustelin *Euterpe* 41-42, and Nohrström 162.
159. For Hultin's biography, see Herman Hultin and Steinby, 162.
160. Reported in Schybergson's 1899 foreword to his edition of Jakob Tengström's writings, the first volume in the series, and Steinby, 56.

161. See Mustelin's 1962 article on this question.
162. *Finsk Tidskrift* 1917, 454-466 (Hultin), 1918, 77-82 (Hedvall) and 82-86 (Hultin).
163. For Homén's biography, see the works by Ekelund and Huldén.
164. See Homén's 1904 *Euterpe* article, which was later collected with his other pieces on the theater in *Från Helsingfors teatrar.*
165. For Boldt's biography, see Nivanka, 74-87.
166. Division minutes, October 14, 1902.
167. For Hasselblatt's life see Steinby, 260-264, and Castrén's 1960 biography.
168. For a brief memorial note on Donner's life, see Nicklin.
169. See Vallinkoski's 1949 article on the history of the dissertation in Finland, 3-4.
170. For Hahl's life, see Koskimies's history of the National Theater, 115-139.
171. Division minutes, December 5, 1896.
172. For Castrén's life, see Barck's 1978 article on him and Fredrik Böök, Enckell, and Steinby, 227-246.
173. Examples of these latter would include "Hjalmar Neiglick som kritiker" ("HN as Critic," 1903), "J. V. Snellmans litteraturkritik och Aleksis Kivi" ("JVS's Literary Criticism and AK," 1920), ("Literary History and Aesthetics," 1925), "Georg Brandes" (1927), "Valfrid Vasenius" (1928), "Henrik Schück" (1935), "Werner Söderhjelm som litteraturkritiker" (1951), "Yrjö Hirn" (1951), and "Sigurd Frosterus som litteraturkritiker" (1958).
174. Cf. Ek, 226.
175. Rafael Koskimies in his article on literary study in the seventh volume of *Suomen kirjallisuus* (Finnish Literature), 252-253.
176. Collected in the 1958 volume *Humanister och humaniora*, 97-104.
177. Gustafsson's voting statement in the division meeting, February 8, 1911.
178. For Mörne's life, see the biographies by Ruin in 1946 and Barck in 1953.
179. For classical studies, see Aalto's 1980 volume, 30-119; for Zilliacus's life, see the same volume by Aalto, 155-157, Landqvist, and Thesleff.
180. For Poirot's work, see Aalto *Modern* 81-83, 218-223.
181. For Tarkiainen's life, see Kari Tarkiainen's biography.
182. [Translator's note. A more idiomatic translation of the "taidekirjailija/kansankirjailija" distinction in contemporary Anglo-American criticism would be "serious/ popular writer." I have rendered Tarkiainen's terms literally in order to capture some of the flavor of Finnish turn-of-the-century cultural politics, specifically the implicit class distinction between the aristocratic/upper-middle-class "art writer" and the lower-middle-class/ working-class/peasant "folk writer."]
183. For a discussion of Tarkiainen's Kivi-image, see also Hein.
184. Among Tarkiainen's students who later defended dissertations showing his methodological influence include Martti Jukola (1900-1952, Ph.D. 1930), Ilmari Heikinheimo (1891-1966, Ph.D. 1933), Kaarlo Nieminen (1887-1955, Ph.D. 1934), Unto Kupiainen (1909-1961, Ph.D. 1939), Irja Maliniemi (1902-1967, Ph.D. 1941), Yrjö Oinonen (1906-1949, Ph.D. 1945), Vilho Suomi (1906-1979, Ph.D. 1952), Eino Kauppinen (1910-, Ph.D. 1952), Ilmari Kohtamäki (1905-1969, Ph.D. 1956), Vihtori Laurila (1904-1988, Ph.D. 1956), and Aarre M. Peltonen (1920-, Ph.D. 1975).
185. For Lönnbohm's life, see Kaukonen.

186. Division minutes May 6, 1895.
187. Division minutes, May 22, 1895.
188. Division minutes, May 22, 1895.
189. Division minutes, December 19, 1900.
190. For a discussion of von Kraemer's work with *Euterpe*, see Mustelin's 1963 book on *Euterpe*, 178-183.
191. Division minutes, October 19, 1912.
192. Division minutes, October 2, 1912.
193. Division minutes, October 2, 1912.
194. For Lehtonen's life, see Koskimies's commemorative oration on him after his death, published in 1950.
195. Recalled by Emma Irene Åström in her 1933 memoirs, 73.
196. See Böök's 1949 study of the two critics, 296-300.
197. Lindberg's year of birth appears as 1869 in the university student register, but in the graduation program for 1897 it is recorded as 1870.
198. See Kannila's edition of Minna Canth's letters, 782, and the University of Helsinki graduation program for 1897, 123-124.
199. Division minutes, May 28, 1915.
200. See Kai Laitinen's 1985 article on Hollo, 200-205.
201. The most important non-Finnish history of Finnish literature in this period was Johann Jakob Meyer's *Vom Land der tausend Seen* (1910), 205 pages of which were literary history, with particular emphasis on late-nineteenth-century and turn-of-the-century literature in Finnish.
202. Reprinted in *Utklipp om böcker*, 401. The Finnish-language literary histories to which Söderhjelm refers are Godenhjelm's *Handbook on the History of Finnish Literature* and J. W. Juvelius's *Suomen kansalliskirjallisuuden vaiheet* ("The Stages of Finnish National Literature").
203. The former was translated into Swedish under the title *Finlands moderna litteratur* in 1968 and into German under the title *Finnlands moderne Literatur* in 1969; the latter appeared in abridged English translation under the title *Literature of Finland* in 1985.
204. For Kallio's life, see Syväoja.
205. Hultin's and Hedvall's polemics appeared in the pages of the *Finsk Tidskrift* for 1917 (454-466) and 1918 (77-86); Söderhjelm's review is reprinted in *Utklipp om böcker*, 399-414.
206. See Rosenberg, 101-116, 146-165.
207. See also Riikonen.

Bibliography

Published and unpublished sources are listed separately. Primary sources are marked with an asterisk.

Abbreviations

AASF	Annales Academiae Scientiarum Fennica
ASSF	Acta Societatis Scientiarum Fennicae
AUU	Acta Universitatis Upsaliensis
FFC	Folklore Fellows Communications
FVSF	Finska Vetenskaps-Societetens förhandlingar (Proceedings of the Finnish Society of Sciences and Letters)
FoU	Förhandlingar och uppsatser (Deliberations and Essays)
HArk	Historiallinen Arkisto (Historical Archives)
HLS	Historiska och litteraturhistoriska studier (Historical and Literary-Historical Studies)
HT	Historiallisia tutkimuksia (Historical Studies)
HYKJ	Helsingin yliopiston kirjaston julkaisuja (Publications of the University of Helsinki Library)
KJ	Kalevalaseuran julkaisuja (Publications of the Kalevala Society)
KTSV	Kirjallisuudentutkijain Seuran vuosikirja (Annual of the Society for Literary Scholars)
KV	Kalevalaseuran Vuosikirja (Kalevala Society Annual)
KVHAA-FFS	Kungliga Vitterhets Historie och Antikvitets Akademien. Filologisk-filosofisk serie (Royal Swedish Academy of Letters, History and Antiquities, Philological/Philosophical Series)
KVST	Kansanvalistusseuran toimituksia (Publications of the Society for Public Education)
MDB	Mitteilungen aus der Deutschen Bibliothek
MLLU	Meddelanden utgivna av Avdelningen för litteratursociologi vid Litteraturhistoriska institutionen i Uppsala (Published Communications of the Divison for the Sociology of Literature at the Department of Literary History in Uppsala)
MSN	Mémoires de la Société Néo-Philologique de Helsinki
MSÅAF	Meddelanden från stiftelsen för Åbo akademi forskningsinstitut (Communications of the Foundation for the Research Institute at Åbo Akademi)
NM	Neuphilologische Mitteilungen

SIJ	Snellman-instituutin julkaisuja (Publications of the Snellman Institute)
SKST	Suomalaisen Kirjallisuuden Seuran toimituksia (Publications of the Finnish Literature Society)
SKHST	Suomen Kirkkohistoriallisen Seuran toimituksia (Publications of the Finnish Church-Historical Society)
SSF	Societas Scientiarum Fennica
SSFU	Skrifter utgivna av Statsvetenskapliga föreningen i Uppsala (Publications of the Political Science Society in Uppsala)
SSLF	Skrifter utgivna av Svenska Litteratursällskapet i Finland (Publications of the Swedish Literature Society in Finland)
STAT	Suomalaisen Tiedeakatemian toimituksia (Publications of the Finnish Adademy of Science)
STAEP	Suomalaisen Tiedeakatemian esitelmät ja pöytäkirjat (Lectures and Minutes of the Finnish Academy of Science)
SÅAK	Skrifter utgivna av Åbo akademi kommitté (Publications of the Åbo Akademi Committee)
TYJ	Turun yliopiston julkaisuja (Publications of the University of Turku)
TYKKJ	Tampereen yliopiston kotimaisen kirjallisuuden julkaisusarja (Publications of the Finnish Literature Department at the University of Tampere)
UAB	Ural-altaische Bibliothek
ÅAÅ	Åbo Akademi årskrift (Åbo Akademi Annual)
ÖFVSF	Öfversigt af Finska Vetenskaps-Societetens förhandlingar (Yearbook of the Finnish Society of Sciences and Letters)

Unpublished Sources

University of Helsinki Senate minutes 1848-1918. University of Helsinki archives.

University of Helsinki Division of History and Linguistics minutes 1853-1918. University of Helsinki archives.

Finnish Literature Society Board of Directors minutes 1879. Finnish Literature Society archives.

Alexander Blomqvist's letters to A. J. Sjögren 1823, 1824, 1848. University of Helsinki Library.

C. G. Estlander's letter to Carl Snoilsky 1883, photocopy. Swedish Literature Society in Finland archives.

Published Sources

Aalto, Pentti. *Classical Studies in Finland 1828-1918.* The History of Learning and Science in Finland 1828-1918. Helsinki, 1980.

–"–. *Modern Language Studies in Finland 1828-1918.* The History of Learning and Science in Finland 1828-1918. Helsinki, 1987.

Ahlqvist, August. *Kalevalan karjalaisuus. Kalevalasta itsestään ja muualta todistettu.* ("The Karelian Origins of the *Kalevala*: Proved by the *Kalevala* Itself and by Other Sources.") Helsinki, 1887.

–"–. Review of Finnish translation of Aristotle's *Poetics. Kieletär* 2(1872):47-51.

*–"–. "Runouden raaka-aine (Esipuhe rohvessori-väitöksessä 1863.)" ("The Raw Materials of Poetry (Lecture at Professorial Dissertation Defense)"). *Koitar* 5(1897):1-10.

*–"–. *Suomalainen runous-oppi kielelliseltä kannalta* ("A Linguistic Approach to Finnish Poetics"). Helsinki, 1863.

*–"–. "Uusin Suomalaisen Virsikirjan ehdotus kielelliseltä ja runolliselta kannalta tarkastettuna. Kutsu kuuntelemaan niitä julkisia esitelmiä . . ." ("The Proposed New Finnish Hymnal Examined from a Linguistic and Poetic Standpoint. An Invitation to Come Hear the Public Lectures . . .") Helsinki, 1884.

*–"–. *Viron nykyisemmästä kirjallisuudesta* ("On Modern Estonian Poetry"). Helsinki, 1855.

Andersin, Hanna. "Kirjailijain muotokuvia XVII. Hippolyte Taine" ("Portraits of Writers XVII: Hippolyte Taine"). *Valvoja* 3/4(1886):153-160, 213-220.

*Anttila, Aarne. *Johannes Linnankosken "Ikuinen taistelu". Kirjallis-historiallinen tutkimus.* ("Johannes Linnankoski's 'The Eternal Battle'") Porvoo, 1922.

–"–. *Elias Lönnrot. Elämä ja toiminta* ("Elias Lönnrot: Life and Work"). 2nd ed. SKST 417. Helsinki, 1985.

Apunen, Osmo. *Hallituksen sanansaattaja Virallinen lehti - Officiella tidningen 1819-1969* ("The Government Messenger *Official Paper* 1819-1969"). Helsinki, 1970.

Arppe, A.E. *Finska Vetenskaps-Societeten 1838-1888. Dess organisation och verksamhet* ("The Finnish Society of Sciences and Letters 1838-1888: Its Organization and Operation"). Helsingfors, 1888.

Aspelin, Eliel. *I anledning af Professor H. Schücks utlåtande om mina vetenskapliga arbeten.* ("On the Occasion of Professor H. Schück's Opinion of My Scholarly Works.") Helsinki, 1900.

–"–. *Kalevalan tutkimuksia.* ("*Kalevala* Studies.") Vol. 1. Helsinki, 1882.

*–"–. *Kansa Saksan kertomarunoudessa 1750-1850.* ("The Folk in German Narrative Literature 1750-1850.") Helsinki, 1894.

–"–. "Kirjallisuudenhistorian tehtävästä." ("On the Function of Literary History.") *Valvoja* 11(1901):619-630.

–"–. "Lamottes afhandlingar om tragedin, granskade och jämförda med Lessing." ("La Motte's Treatments of Tragedy, Examined in Comparison with Lessing.") ASSF 16(1886):1-60.

*–"–. *Lars Stenbäck.* Helsinki, 1900.

–"–. "Runoilija Alexis Kiwi." ("The Poet Aleksis Kivi.") *Kirjallinen Kuukauslehti* 10/11(1872):227-240.

–"–. "Vieläkin 'estetiikasta.'" ("Once Again on Aesthetics.") *Valvoja* 2(1895):120-123.

–"–. "Yleinen kirjallisuushistoria suomeksi." ("A General History of Literature in Finnish.") *Uusi Suometar* (August 12, 1886).

Aspelin-Haapkylä, Eliel. "Aleksis Kiven 'Seitsemän Veljestä.'" ("On Aleksis Kivi's *Seven Brothers*.") *Aika* 1(1911):1-15, 4(1911):225-231.

–"–. *Suomalaisen Teatterin historia. I-IV.* ("History of the Finnish Theater, 4 vols.) SKST 115. Helsinki, 1906, 1907, 1909, 1910.

–"–. *Kirovuosien kronikka. Otteita professori Eliel Aspelin-Haapkylän päiväkirjasta vuosilta 1905-1917.* ("A Chronicle of the Plague Years: Selections from Professor Eliel Aspelin-Haapkylä's Journal, 1905-1917.") Edited by Eero Saarenheimo. SKST 357. Helsinki, 1980.

*–"–. *Muoto- ja muistikuvia. I-III.* ("Portraits and Memories." 3 vols.) Helsinki, 1911, 1911, 1914.

Autio, Veli-Matti. *Yliopiston virkanimitykset 1809-1852. Hallinto- ja oppihistoriallinen tutkimus Turun Akatemian ja Keisarillisen Aleksanterin-Yliopiston opettajien nimityksistä Venäjän vallan alkupuolella.* ("University Appointments 1809-1852: A Study in the History of Administration and Learning of Faculty Appointments at the Turku Academy and the Czar Alexander University During Russian Rule.") HT 115. Helsinki, 1981.

Barck, P.O. *Arvid Mörne och sekelskiftets Finland.* ("Arvid Mörne and Turn-of-the-Century Finland.") Helsingfors, 1953.

–"–. "Gunnar Castrén och [and] Fredrik Böök." HLS 53. SSLF 478. 249-287. Helsinki, 1978.

Bennich-Björkman, Bo. *Termen litteratur i svenskan 1750-1850.* ("The Word 'Literature' in Swedish 1750-1850.") MLLU 4. Uppsala, 1970.

*Bergbom, Karl Johan. *Det historiska dramat i Tyskland.* ("The Historical Drama in Germany.") Helsinki, 1868.

Bergbom, Kaarlo. *Kirjoitukset. II. Tutkimukset ja arvostelut.* ("Criticism and Reviews. Vol. 2 of *Writings.*") Helsinki, 1908.

*–"–. Review of C. G. Estlander's *De bildande konsternas historia ifrån slutet af första århundradet intill våra dagar. Kirjallinen Kuukauslehti* 1(1869):13-15.

*Bergholm, A. H. *Abraham Poppius. Elämäkerta ja runot.* ("Abraham Poppius: His Life and Poetry.") *Suomi* 3.17. Helsinki, 1899. 1-75.

–"–. "Sananen kirjallishistoriamme kirjoittamisesta." ("A Word About the Writing of Our Literary History.") *Tellervo* 4(1910):98-103.

*–"–. *Studier öfver C.J.L. Almqvist.* ("Studies of C. J. L. Almqvist.") Helsinki, 1902.

*Berndtson, Fridericus. *Adumbratio notionis phantasiae pulchrae.* Helsinki, 1847.

*–"–. *De Vi et Essentia ingenii in arte se manifestantis.* Helsinki, 1847.

*Berndtson, Fredrik. *Dramatiska studier och kritiker.* ("Dramatic Studies and Reviews.") Helsinki, 1879.

*–"–. "F. M. Franzén." In Berndtson, *Litterära Soiréer i Helsingfors under hösten 1849,* 3:1-23.

–"–. *Litterära Soiréer i Helsingfors under hösten 1849.* ("Literary Soirees in Helsinki in Autumn 1849.") Helsinki, 1849.

"Biographi öfver framledne theol. doctorn, prosten och kyrkoherden i Sotkamo, herr magister Johan Frosterus." ("A Biography of Johan Frosterus, D.D., Minister and Vicar in Sotkamo.") *Mnemosyne* 60-69(1820):237-274.

Biografinen nimikirja. Elämäkertoja Suomen entisiltä ja nykyajoilta. ("Biographical Register: Biographies from Former and Modern Times in Finland.") Edited by the Finnish Historical Society. Helsinki, 1879-1883, 1889.

*Björkenheim, Magnus. *Äldre fransk litteratur på herrgårdar i Finland.* ("Older French Literature in Finnish Manor Houses.") Helsinki, 1929.

Björkstén, Carita. "Fredrik Berndtson." In vol. 2 of *Nyländska öden.* ("Uusimaa Destinies.") 47-104. Tammisaari, 1966.

*Blomqvist, Alexander. *De principio methodi ethnographicae in historia litteraria universali adhibendae schediasma.* Helsinki, 1838.

*–"–. *Quae ab Aeschylo accesserunt momenta tragediae Graecae c. IV.* Turku, 1823.

*Boldt, Alexander. *Erik Gustav Geijer. En levnadsteckning med utdrag ur Geijers brev och vittra skrifter. I-III.* ("Erik Gustav Geijer: A Biography with Excerpts from Geijer's Letters and Literary Writings." 3 vols.) Helsinki, 1909, 1910.

*–"–. *Studier öfver en del barndomen och ynglingaåldern berörande drag i svenska och svensk-finska skalders diktning. I.* ("Studies of the Treatment of Childhood and Adolescence in the Poetry of Some Swedish and Swedish-speaking Finnish Poets.") Helsinki, 1902.

Bolin, Wilhelm. Review of Henrik Schück, *William Shakespeare. Finsk Tidskrift* 21(1886):187-199.

–"–. Review of Robert Prölss, *Katechismus der Aesthetik. Finsk Tidskrift* 6(1879):165-166.

–"–. Review of Valfrid Vasenius, *Henrik Ibsen, ett skaldeporträtt.* ("Henrik Ibsen: Portrait of a Poet.") *Finsk Tidskrift* 15(1883):6-9.

*Borenius, A. A. "Missä Kalevala on syntynyt?" ("Where Did the *Kalevala* Originate?") *Suomen Kuvalehti* 23(1873):269-274.

Borg, C. G. *Mathias Alexander Castrén. Uppläst vid Österbottniska afdelningens årsfest den 9 nov. 1852.* ("Mathias Alexander Castrén: Recited at the Annual Festivities of the Ostrobothnian Division, November 9, 1852.") Helsinki, 1853.

*Bremer, K. V. *Kurs i svenska literaturens historia för fruntimmersskolan i Hfors.* ("Course in the History of Swedish Literature for the Women's School in Helsinki.") Helsinki, 1874.

Bure, Anders. *Pohjoismaiden kuvaus v. 1626.* ("Depiction of the Nordic Countries in 1626.") Trans. and annotated, and with an introduction, by Tuomo Pekkanen. SKST 427. Mikkeli, 1985.

Böök, Fredrik. *Essayer och kritiker 1917-1918.* ("Essays and Reviews.") Stockholm, 1919.

–"–. *Victoria Benedictsson och [and] Georg Brandes.* Stockholm, 1949.

Cajanus, Johannes. "Biographi öfver mag. Johannes Cajanus, extra-ord. professor vid Universitetet i Åbo." ("Biography of M.A. Johannes Cajanus, professor extraordinaire at the University of Turku.") *Mnemosyne* 21(1819):81-84.

*Calamnius, Johan Wiktor. *Försök till framställning af Aristoteles' lära om det sköna och konsten.* ("Inquiry into the Presentation of Aristotle's Theory of the Beautiful and Art.") Helsinki, 1865.

Canth, Minna. *Minna Canthin kirjeet.* ("The Letters of Minna Canth.") Edited by Helle Kannila. SKST 313. Vammala, 1973.

Carpelan, Tor, ed. *Finsk biografisk handbok.* ("Finnish Biographical Handbook.") 2 vols. Helsinki, 1903.

Carpelan, Tor, and L. O. Th. Tudeer, eds. *Helsingin yliopiston opettajat ja virkamiehet vuodesta 1828.* ("Faculty and Administrators at the University of Helsinki since 1828.") 2 vols. Helsinki, 1925.

*Castrén, Gunnar. "Emil Hasselblatt." HLS 35. SSLF 375. 5-42. Helsinki, 1960.

*–"–. *Frans Michael Franzén i [in] Finland.* Helsinki, 1902.

*–"–. *G. Bernard Shaw.* Konst och kultur ("Art and Culture") 4. Helsinki, 1906.

*–"–. *Gustav Philip Creutz.* Helsinki, 1917.

–"–. "Helsingfors universitet och den skandinaviska kulturen." ("The University of Helsinki and Scandinavian Culture." *Hufvudstadsbladet* (August 19, 1940).

*–"–. *Herman Kellgren. Ett bidrag till 1840- och 1850-talens kulturhistoria.* ("Herman Kellgren: A Contribution to the Cultural History of the 1840s and 1850s.") SSLF 302. Helsinki, 1945.

–"–. *Humanister och humaniora. Tryckt och talat från sex decennier.* ("Humanists and the Humanities: Words Written and Spoken Over Six Decades.") Helsinki, 1958.

*–"–. *Juhani Aho.* 2 vols. Helsinki, 1922.

*–"–. *Norden i den franska litteraturen.* ("The Nordic Countries in French Literature.") Helsinki, 1910.

*–"–. *Runar Schildt.* Helsinki, 1927.

–"–. "Runeberg och universitetet. Några anteckningar." ("Runeberg and the University: Some Notes.") HLS 12. SSLF 254. 133-148. Helsinki, 1936.

*–"–. *Stormaktstidens diktning. Studier.* ("The Poetry of the Great Power Period: Studies.") Helsinki, 1907.

–"–. "Svenskt i Finland." ("Swedish in Finland.") *Nya Argus* 11(1914):96-99.

*Cederberg, Eino. *Tuhlaajapoika-aihe. Sen pohjoismaiset edustajat ja sen suhde ylioppilaskomedioihin. Tutkimuksia koulunäytelmän alalta.* ("The Prodigal Son Motif: Its Nordic Representatives and its Connections with Student Comedies; Studies of School Plays.") Helsinki, 1917.

Chydenius, Anders. Review of Gabriel Lagus, *Den Finsk-Svenska Litteraturens Utveckling I-II. Helsingfors Dagblad* (November 18, 1867).

*Clasen, Alexis Theodorus. *De Galfredo Chaucero poeta statuque ante eum poesis anglicae.* Helsinki, 1851.

*Collan, Karl. *Königinhofer- och Grünbergerhandskriften. Från fornböhmiskan öfversatta, jemte litterärhistorisk inledning och upplysande noter.* ("The Könighofer and Grünberger Manuscripts: Translated from the Old Bohemian, with a Literary-Historical Introduction and Informative Notes.") Helsinki, 1865.

–"–. "Om [on] 'Das Nibelungenlied.'" *Litteraturblad* 1(1858):7-24.

*–"–. *Studier och skizzer.* ("Studies and Sketches.") Helsinki, 1865.

*–"–. *Öfversigt af Serviens historiska folksånger.* ("Overview of Serbian Folk Poetry.") Helsinki, 1860.

Colliander, O. I. *Suomen kirkon paimenmuisto 19:n vuosisadan alusta nykyaikaan.* ("The Pastoral Letter in the Finnish Church from the Early Nineteenth Century to the Present.") Vol. 2. Helsinki, 1918.

*Cygnaeus, Fredrik. *Afhandlingar i populära ämnen.* ("Disquisitions on Popular Themes.") 2 vols. Helsinki, 1853.

*–"–. "Det tragiska elementet i Kalevala." ("The Tragic Element in the *Kalevala.*") In Cygnaeus, *Afhandlingar i populära ämnen.* 2:69-131.

–"–. "En [a] 'lectio praecursoria.'" *Litteraturblad* 5(1856):148-155.

*–"–. *Erik XIV såsom dramatisk karakter.* ("Erik XIV as Dramatic Character.") Helsinki, 1853.

–"–. "Ett och annat angående Elgskyttarne och Hanna, dikter af Joh. Ludv. Runeberg." ("This and That About the *Elk Hunters* and *Hanna* by J. L. Runeberg.") In Cygnaeus, *Jääkynttilät, ströskrift i fria häften.* ("Ice Candles: Free-Bound Pamphlets.") Vol. 1, 7-107. Helsinki, 1837.

*–"–. *Johan Jacob Nervander.* Helsinki, 1846.

–"–. "Några ord i anledning af underrättelsen om Joh. Robert Tengströms bortgång, talade inför den akademiska läseföreningens medlemmar i början af December 1847." ("Some Words upon the Decease of J. Robert Tengström, Spoken Before the Members of the Academic Reading Group in Early December, 1847.") In Cygnaeus, *Afhandlingar i populära ämnen,* 2:14-33.

*–"–. *Om "Fänrik Ståls sägner". Betraktelser.* ("On *The Songs of Ensign Stal:* Reflections.") Helsinki, 1861.

*–"–. "Om Nadeschda." ("On [Runeberg's] *Nadeschda.*") *Helsingfors Morgonblad* (May 17/24, 1841).

*–"–. Review of Aleksis Kivi, *Nummisuutarit* (*Heath Cobblers*). *Helsingfors Tidningar* March 7/27, April 11, 1865.

–"–. *Små häften angående litteratur och konst.* ("Thin Volumes on Literature and Art.") Vol. 2. Helsinki, 1868.

–"–. *Teckningar ur F. M. Franzéns lefnad vid åminnelsefesten den 9 februari 1872 i Finska Universitetets högtidssal.* "Notes on the Life of F. M. Franzén in Commemoration of his Birth, February 9, 1872, in the Finnish University Stateroom.") Helsinki, 1872.

*–"–. *Om [on] Johan Ludvig Runeberg.* Helsinki, 1873.

*Donner, J. O. E. *Der Einfluss Wilhelm Meisters auf den Roman der Romantiker.* Helsinki, 1893.

*–"–. *Lord Byrons Weltanschauung.* ASSF 22:4. Helsinki, 1897.

*Donner, Otto. "Kalevipoeg jumalaistarulliselta ja historialliselta kannalta katsottuna." ("Kalevipoeg as Myth and History.") *Suomi* 2.5(1866):145-207.

*Edlund, Erik. *M. A. Castrénin elämä ja matkustukset, nuorisolle kerrotut.* ("The Life and Travels of M. A. Castrén, Narrated for Young People.") Helsinki, 1872.

Edman, Alfred. *Hegels lära om konstformerna.* ("Hegel's Teaching on Forms of Art.") Helsinki, 1856.

Ek, Sverker. *Franzéns Åbodiktning. Ett bidrag till Finlands svenska litteraturhistoria.* ("Franzén's Turku Poetry: a Contribution to the History of Finnish Literature Written in Swedish.") Stockholm, 1916.

Ekelund, Erik. "Olaf Homén." *Åbo Akademi Annual 1949-1951,* 75-79. Turku, 1952.

Elmgren, S. G. "Berättelse om Finska litteratursällskapet under de första 17 åren." ("Report on the Finnish Literature Society in Helsinki During the First 17 Years.") *Suomi* 1.7 (1847). Helsinki, 1848. 1-45.

–"–. *S. G. Elmgrenin muistiinpanot. Suomen historian lähteitä II.* (The Notes of S. G. Elmgren: Sources of Finnish History," vol. 2.) Edited by Aarno Maliniemi. Helsinki, 1939.

*–"–. *Öfversigt af Finlands litteratur.* ("Overview of Finnish Literature.") Vol. 1, *Ifrån år 1542 till 1770.* ("From 1542 to 1770.") Vol 2, *Ifrån år 1771 till 1863.* ("From 1771 to 1863.") Helsinki, 1861, 1865.

Enckell, Olof. "Gunnar Castrén. Minnestal." ("Gunnar Castrén: a Commemorative Address.") *Societas Scientiarum Fennica Årsbok/Vuosikirja* (Annual) 39 C 3. 3-14. Helsinki, 1962.

*Erich, Mikko W. *Ugo Foscolo come uomo e come poeta lirico.* ("Ugo Foscolo as Man and Poet.") Florence, 1912.

*–"–. *William Shakespeare ja hänen runoutensa.* ("William Shakespeare and His Poetry.") Helsinki, 1916.

*Estlander. C. G. *Bernardin de Saint-Pierre och hans mémoire om Finland.* ("Bernardin de Saint-Pierre and His Memoirs of Finland.") Academic Invitation. Helsinki, 1898.

*–"–. *Bidrag till den provencaliska litteraturens historia.* ("A Contribution to the History of Provencal Literature.") Helsinki, 1868.

*–"–. "Det estetiska och moraliska intresset i skaldekonsten." ("The Aesthetic and Moral Interest in Poetic Art.") *Finsk Tidskrift* 13(1882):198-216.

*–"–. *Folksångerna om Robin Hood.* ("Folk Songs About Robin Hood.") Helsinki, 1859.

–"–. *Från flydda tider. Uppsatser, skildringar och tal.* ("From Bygone Times: Essays, Accounts, and Talks.") Helsinki, 1905.

*–"–. *J. L. Runebergs estetiska åsigter.* ("J. L. Runeberg's Aesthetic Views.") Helsinki, 1888.

–"–. "Mistral och Mirèio. Minnen och bedömanden." ("Mistral and Mirèio: Recollections and Judgments.") *Finsk Tidskrift* 58(1905):455-473.

*–"–. *Naturalismen enligt Zola.* ("Naturalism According to Zola.") Academic Invitation. Helsinki, 1891.

*–"–. "Om förhållandet mellan konsten och moralen." ("On the Relation Between Art and Morality.") 1894.

*–"–. "Pièces inédites du roman de Tristan, précédées de recherches sur son origine et son développement." ASSF 8:2, 413-449. Helsinki, 1867.

*–"–. *Poema del Cid i svensk öfversättning med historisk och kritisk inledning.* ("The Poem del Cid in Swedish Translation, with a Historical and Critical Introduction.") Helsinki, 1863.

–"–. Review of C. R. Nyblom's edition of *Johan Ludvig Runebergs Samlade skrifter* ("Collected Works"). *Helsingfors Dagblad* (October 10, 1871).

–"–. Review of Fredrik Cygnaeus, *Samlade arbeten* ("Collected Works"). *Finsk Tidskrift* 34(1893):245-258, 325-337.

–"–. Review of Valfrid Vasenius, *Henrik Ibsens dramatiska diktning i dess första skede. Finsk Tidskrift* 9(1880):221-224.

–"–. Review of Wilhelm Lagus, *Skalden Johan Henrik Kellgrens finska lefnadsminnen. Finsk Tidskrift* 18(1885):57-60.

*–"–. *Richard Lejonhjerta i historien och poesin.* ("Richard the Lionhearted in History and Poetry.") Helsinki, 1858.

*–"–. "J. L. Runebergs religiösa verldsåskådning." ("Runeberg's Religous Outlook.") FoU 1:1-14. SSLF 2. Helsinki, 1886.

*–"–. *Runebergs skaldskap. Kritiska studier föregångna af redogörelse för normalupplagans redaktion samt en kronologisk förteckning.* ("Runeberg's Poetry: Critical Studies Preliminary to an Outline for the Variorum and Chronological Catalogue of a Normal Edition.") Helsinki, 1902.

–"–. "'Svartsjukans Nätter' af Runeberg." ("Runeberg's 'Jealous Nights.'") *Litteraturblad* 2(1857):46-56.

*–"–. "Till läsaren." ("To the Reader.") *Finsk Tidskrift* 1:1876)3-16.

–"–. *Ungdomsminnen.* ("Recollections of Youth.") Helsinki, 1918.

–"–. *Uppsatser i litteratur och konst samt allmänna ämnen. 2. Åren 1882-1893.* ("Essays on Literature and Art, and General Subjects. 2. 1882-1893.") Vol. 2 of *Skrifter* ("Writings"). SSLF 146. Helsinki, 1919.

–"–. *Uppsatser om Johan Ludvig Runeberg.* ("Essays on J. L. Runeberg.") Vol. 1 of *Skrifter* ("Writings"). SSLF 117. Helsinki, 1914.

–"–. *Vid konstflitens härdar i Tyskland, Österrike, Schweiz och Belgien. Reseanteckningar.* ("In the Industrial Art Centers of Germany, Austria, Switzerland, and Belgium: Travel Notes.") Helsinki, 1875.

*–"–. *Vitterhetens utveckling hos de nyare folken.* ("Literary Development Among the Modern Peoples.") Vol. 1. *Medeltiden. Förra perioden.* ("Middle Ages: Earlier Period.") Helsinki, 1900.

Finlands minnesvärde män. Samling af lefnadsteckningar. ("Finland's Notable Men: A Collection of Biographies.") 2 vols. Helsinki, 1853, 1855-1857.

"Finska språket och Finska Vetenskaps-Societeten." ("The Finnish Language and the Finnish Society of Sciences and Letters.") *Morgonbladet* (November 12, 1849).

Forselles, Jenny af. *A. N. Clewberg-Edelcrantz och hans omgifning.* ("A. N. Clewberg-Edelcrantz and His Environment.") Helsinki, 1903.

Forsman, A. V. Review of Julius Krohn, *Suomalaisen kirjallisuuden vaiheet. Virittäjä* 2(1898):17-27.

Forsman [Koskimies], Rafael. *Fredrik Cygnaeus kirjailijana ja ajanilmiönä.* ("Fredrik Cygnaeus as Writer and Product of His Times.") 2 vols. Helsinki, 1923, 1925.

*Fraser, Georg. *Vitterheten under nittonde seklets reformtid i Ryssland.* ("Literature in the Nineteenth-Century Reform Period in Russia.") Helsinki, 1888.

Fromm, Hans. "Zur Rezeptionsgeschichte des Kalevala." Congressus Quintus Internationalis Fennougristarum. 25-55. Turku, 1980.

*Geitlin, Gabriel. *De meritis litterariis Lomonossovii.* 4 vols. Helsinki, 1829, 1830.

*Godenhjelm, B. F. *Lyhyt runous-oppi. Jatko-opistoja varten.* ("A Brief Poetics: For Teachers Training Colleges.") Helsinki, 1891.

*–"–. *Oppikirja suomalaisen kirjallisuuden historiassa.* ("Textbook on the History of Finnish Literature." Helsinki, 1884. Trans. by E. D. Butler as *Handbook of the History of the Finnish Literature.* London, 1896.

–"–. Review of E. W. Palander, *Uebersicht der neueren russischen Literatur. Kirjallinen Kuukauslehti* 10(1880):234-236.

*–"–. *Runous ja runouden muodot. Kirjoitelmia, runoja.* ("Poetry and Its Forms: Writings and Poems.") SKST 135. Helsinki, 1914.

Gottlund, C. A. "Sampo, ja muutama sana nykyisestä Runollisuutestamme." ("The Sampo, and a Word or Two About Our Current Poetry.") *Suomi* (July 3, 1847).

Grimm, Jacob. "Über das finnische Epos." 1845. Trans. into Swedish as *Om det Finska Epos. Fosterländsk Album* (1845): 60-103.

Grotenfelt, Arvi. "Historiallinen yhtenäisyys kansalliskirjallisuudessa." ("Historical Unity in National Literature.") *Valvoja* 3(1909):161-163.

Grotenfelt, Kustavi. "Elämäkerta." ("Biography [of B. F. Godenhjelm].") In B. F. Godenhjelm, *Runous ja runouden muodot.*

–"–. Review of Julius Krohn, *Suomalaisen kirjallisuuden vaiheet. Valvoja* 2(1898):133-137.

Grönroos, Henrik. "Mitä kirjat kertoivat Saksasta ja Venäjästä 1700-luvun suomalaisille." ("What Did Eighteenth-Century Finns Read About Germany and Russia in Books?") *Miscellanea bibliographica* 4, 62-84. Helsinki, 1944.

Gustafsson, Lars. *Estetik i förvandling. Estetik och litteraturhistoria i Uppsala från P. D. A. Atterbom till B. E. Malmström.* ("Aesthetics in Transition: Aesthetics and literary history in Uppsala from P. D. A. Atterbom to B. E. Malmström.") AUU, Historia litterarum, 16. Uppsala, 1986.

–"–. *Litteraturhistorikern Schück. Vetenskapssyn och historieuppfattning i Henrik Schücks tidigare produktion.* ("Literary Historian Schück: The Conception of Science and History in the Early Work of Henrik Schück.") AUU, Historia litterarum, 12. Uppsala, 1983.

Gyldén, Nils Abraham. *Till åhörande af det offentliga föredrag, hvarmed professoren i Estetik och nyare litteratur Mag. Fredrik Cygnaeus kommer att tillträda sitt ämbete.* ("To the Audience of the Open Lecture by Professor of Aesthetics and Modern Literature Fredrik Cygnaeus upon his Inauguration into Office.") Helsinki, 1854.

Haapanen, Toivo. *Suomalaiset runomittateoriat 1800-luvulla.* ("Finnish Theories of Poetic Meter in the Nineteenth Century.") Vol. 1 of *Suomi* 5.6. Helsinki, 1926.

Haavio, Martti. "Kalevalakultti." ("The *Kalevala* Cult.") In F. A. Heporauta and Martti Haavio (eds.), *Kalevala, kansallinen aarre.* ("*Kalevala*, the National Treasure.") Porvoo, 1949.

*Hagman, Lucina. *Fredrika Bremer. Kuvaus vuosisatamme alkupuolelta.* ("Fredrika Bremer: A Portrait of the Early Years of Our Century.") Porvoo, 1886.

*Hagman, Lucina. *Minna Canthin elämäkerta.* ("A Biography of Minna Canth.") 2 vols. Kuopio, Helsinki, 1906, 1911.

*Hahl, Hjalmar. *Les tendances morales dans l'oeuvre de Giacomo Leopardi.* Helsinki, 1896.

*Hahl, Jalmari. *Nykyajan näytelmäkirjoittajia.* ("Modern Playwrights.") Helsinki, 1908.

Haltsonen, Sulo. "Serbialainen kansanrunous Suomessa." ("Serbian Folk Poetry in Finland.") KTSV 12, 57-67. Helsinki, 1952.

–"–. ed. "Luettelo suomalaisista kirjallisuudentutkimuksista 1901-1925." ("Index of Finnish Literary Research, 1901-1925.") KTSV 11 (1954).

Hannikainen, Pietari. "Suomalainen kirjallisuus wuonna 1846." ("Finnish Literature in 1846.") *Kanava* (January 20, 1847).

*Hasselblatt, Emil. *Dikt och diktare. Studier och kritiker.* ("Poem and Poets: Studies and Reviews.") Helsinki, 1918.

Hautala, Jouko. *Finnish Folklore Research 1828-1918.* Trans. Eija Kennerley, Peter Jones, and Kingsley Hart. The History of Learning and Science in Finland 1828-1918. Helsinki, 1968.

–"–. *Suomalainen kansanrunoudentutkimus.* ("Finnish Folklore Research.") SKST 244. Turku, 1954.

*Havu, I. *Pietari Päivärinta. Kirjallishistoriallinen tutkimus.* ("Pietari Päivärinta: A Literary-Historical Study.") Porvoo, 1921.

*Hedvall, Ruth. *Finlands svenska litteratur.* ("The Swedish-Language Literature of Finland.") SÅAK 5. Porvoo, 1917.

*–"–. *Friedrich Hölderlin. En studie.* ("Friedrich Hölderlin: A Study.") Helsinki, 1921.

*–"–. *Johan Ludvig Runeberg och hans diktning.* ("J. L. Runeberg and His Poetry.") Tampere, 1931.

*–"–. "Modern stilforskning." ("Modern Stylistics.") *Finsk Tidskrift* 79(1915):176-183.

*–"–. *Runebergs poetiska stil. Studier.* ("Runeberg's Poetic Style: Studies.") Helsinki, 1915.

Heikel, I. A. *Filologin.* ("Philology.") Vol. 5 of *Åbo universitets lärdomshistoria.* ("The History of Learning at the University of Turku.") SSLF 26. Helsinki, 1894.

Hein, Manfred Peter. *Kanonisierung eines Romans. Alexis Kivis "Sieben Brüder" 1870-1980.* Keuruu, 1984.

"Henrik Gabriel Porthan." Obituary, *Åbo Tidning* March 17, 1804.

Henrikson, Thomas. *Romantik och marxism. Estetik och politik hos Otto Ville Kuusinen och Diktonius till och med 1921.* ("Romanticism and Marxism: Aesthetics and Politics in the Work of Otto Ville Kuusinen and [Elmer] Diktonius up to and including 1921.") Helsinki, 1971.

Hipping, A. J. *Åminnelse-tal öfver framlidne erkepiskopen, herr doctor Jacob Tengström, hållet i Åbo gymnasii solennitetssal d. 13 junii 1842.* "Commemorative Address on the Late Archbishop, Dr. Jacob Tengström, Delivered in the Turku High School Auditorium, July 13, 1842.") Turku, 1843.

*Hirn, Yrjö. *Barnlek. Några kapitel om visor, danser och små teatrar.* ("Children's Play: Some Essays on Songs, Dances, and Plays.") Helsinki, 1916. Trans. into French by T. Hammar as *Les jeux d'enfants.* Paris 1926. Trans. into Italian by Paola Faggioli as *I giuochi dei bimbi.* Venice, 1929.

–"–. *Beaumarchais. Äventyrare, stridsman och författare.* (Beaumarchais: Adventurer, Soldier, and Writer.") Helsinki, 1931.

*–"–. *Den gamla postvagnen och några av dess passagerare.* ("The Old Stagecoach and Some of its Passengers.") Helsinki, 1926.

*–"–. *Det estetiska lifvet.* ("The Aesthetic Life.") Stockholm, 1913.

*–"–. *Det heliga skrinet. Studier i den katolska kyrkans poesi och konst.* Stockholm, 1909. Trans. into English as *The Sacred Shrine: A Study of the Poetry and Art of the Catholic Church.* London, 1912.

*–"–. *Diderot.* Helsinki, 1917.

*–"–. *Dr Johnson och James Boswell. En bok om engelskt liv och lynne.* ("Dr. Johnson and James Boswell: A Book on English Life and Character.") Helsinki, 1922.
*–"–. *Episoder. Några kapitel om lif och litteratur under sjuttonhundratalet.* ("Episodes: Some Essays on Life and Literature in the Eighteenth Century.") Helsinki, 1918.
*–"–. *Eremiter och pilgrimer.* ("Hermits and Pilgrims.") Helsinki, 1924.
*–"–. *Förstudier till en konstfilosofi på psykologisk grundval.* ("Prolegomena for a Philosophy of Art Grounded in Psychology.") Helsinki, 1896.
*–"–. *Goda vildar och ädla rövare i dikt och verklighet.* ("The Noble Savage and Honorable Thief in Poetry and Reality.") Helsinki, 1941.
*–"–. *Jonathan Swift.* Helsinki, 1918.
*–"–. *Konsten och den estetiska betraktelsen. Tre föreläsningar, hållna vid Londons universitet i maj 1936.* ("Art and Aesthetic Reflection: Three Lectures Delivered at the University of London in May 1936.") Helsinki, 1937.
*–"–. *Lärt folk och landstrykare i det finska Finlands kulturliv.* ("Learned Folk and Vagrants in Finnish Cultural Life.") Helsinki, 1939.
–"–. "Nya arbeten i litteraturhistoria och biografi." ("New Works in Literary History and Biography.") *Nya Argus* 24(1915):225-227.
*–"–. *The Origins of Art: A Psychological and Sociological Inquiry.* London, 1900. Trans. into Swedish as *Konstens ursprung. En studie öfver den estetiska värksamhetens psykologi och sociologiska orsaker.* Helsinki, 1902. Trans. into German by M. Barth as *Der Ursprung der Kunst. Eine Untersuchung ihrer psychischen und sozialen Ursachen.* Leipzig, 1904. Trans. into Russian as *Proishozdjeniy iskusstva.* Harjkov, 1923.
*–"–. *Runeberggestalten. Strödda studier.* ("The Figure of Runeberg: Casual Studies.") Helsinki, 1942.
*–"–. *Runebergskulten.* ("The Runeberg Cult.") SSLF 248. Helsinki, 1935.
–"–. "Svenska Litteratursällskapets nya edition av Strömborgs Runebergbiografi." ("The Swedish Literature Society's New Edition of Strömborg's Biography of Runeberg.") *Finsk Tidskrift* 112(1932):28-42.
–"–. *Teatrar och teaterstrider i 1800-talets Finland.* ("Theaters and Their Struggles in Nineteenth-Century Finland.") Helsinki, 1949.
*–"–. "Vertailevien taidetutkimusten metoodeista." ("The Methods of Comparative Art Research.") *Valvoja* 7(1897):446-455.
*–"–. *Ön i världshavet.* ("The Island in the Ocean.") Helsinki, 1928.
*Hollo, J. A. "Huomioita Aleksis Kiven lyriikasta." ("Observations on Aleksis Kivi's Poetry.") *Valvoja* 8(1912):556-567.
*–"–. "Sananen Aleksis Kiven lyriikasta." ("A Word About Aleksis Kivi's Poetry.") *Suomalainen* 2(1911):132-138.
*Holsti, R. I. "Om Hamlet af Shakespeare." ("On Shakespeare's Hamlet.") *Joukahainen* 2(1845):44-88.
*Homén, Olaf. *De nya författarna.* ("Recent Writers.") Porvoo, 1915.
–"–. "En glömd publicist. Carl Wilhelm Törnegren." ("A Neglected Publicist: Carl Wilhelm Törnegren.") *Lucifer* 3(1918):58-64.
–"–. "Ett nytt teaterföretag på inhemsk grund." ("A New Theatrical Undertaking on Native Grounds.") *Euterpe* 9(1904):102-104.

*–"–. *Från Helsingfors teatrar.* ("Helsinki Theaters.") 3 vols. Helsinki, 1915, 1918, 1919.
–"–. *Gabriel Lagus.* Vol. 1. Helsinki, 1921.
*–"–. *Poetik.* ("Poetics.") Revised and published by Erik Ekelund. Helsinki, 1953.
*–"–. *Studier i fransk klassicism 1630-1665.* ("Studies in French Classicism.") 2 vols. Helsinki, 1914, 1916.
Honko, Lauri. "Kalevala: aitouden, tulkinnan ja identiteetin ongelmia." ("The *Kalevala*: Problems of Authenticity, Interpretation, and Identity.") In Lauri Honko ed., *Kalevala ja maailman eepokset.* ("*Kalevala* and World Epics.") KV 65, 125-170. Helsinki, 1987.
Huldén, J. J. "Olaf Homén." *Svensk litteraturtidskrift* 4(1950):145-158.
*Hultin, Arvid. "Bidrag till Litteratur- och Lärdomshistorien i Finland. Några anteckningar från perioden 1640-1720." ("A Contribution to the History of Literature and Learning in Finland: Some Notes on the Period 1640-1720.") FoU 27, 1-66. SSLF 115. Helsinki, 1914.
*–"–. *Den svenska vitterheten i Finland under stormaktstiden 1640-1720. Valda vitterhetsarbeten jämte en litteraturhistorisk inledning.* ("Swedish Literature in Finland During the Great Power Era, 1640-1720: Selected Literary Works and a Literary-Historical Introduction.") SSLF 65. Helsinki, 1904.
*–"–. *Det ekonomiska tidevarvet i Finlands litteraturhistoria. Ur odlingens hävder under frihetstidens senare del.* ("The Economic Period in the History of Finnish Literature: The Annals of Culture in the Later Years of the Age of Freedom.") SSLF 94. Helsinki, 1910.
–"–. "En finländsk dramatiker på 1830-talet." ("A Finnish Playwright in the 1830s.") FoU 4:99-120. SSLF 13. Helsinki, 1889.
*–"–. "En finsk vitterlekare i slutet af 17:de seklet." ("A Finnish Writer at the End of the Eighteenth Century.") FoU 9 (257-309) and 10 (194). SSLF 30, 34. Helsinki, 1895.
*–"–. *Finlands litteratur under frihetstiden.* Vol. 1 ("Finnish Literature During the Age of Freedom.") Helsinki, 1906.
–"–. "Frihetstidens Litteratur- och Lärdomshistoria i Finland." ("The History of Literature and Learning in the Age of Freedom in Finland.") FoU 28, 229-313. SSLF 121. Helsinki, 1915.
*–"–. *Gustaf Filip Creutz. Hans levnad och vittra skrifter.* ("Gustaf Filip Creutz: His Life and Literary Writings.") SSLF 111. Helsinki, 1913.
–"–. *Litteraturstudier i Åbo under Porthans tid.* ("Literary Studies in Turku During Porthan's Time.") FoU 31, 1-84. SSLF 140. Helsinki, 1918.
–"–. *Litteraturstudier i Åbo under svenska tiden 1640-1809.* ("Literary Studies in Turku in the Swedish Era, 1640-1809.") Helsinki, 1918.
–"–. Review of Ruth Hedvall, *Finlands svenska litteratur. Finsk Tidskrift* 83(1917):454-466.
*–"–. *Torsten Rudeen. Ett bidrag till karolinska tidens litteratur- och lärdomshistoria.* ("Torsten Rudeen: A Contribution to the History of Literature and Learning in the Carolian Period.") SSLF 54. Helsinki, 1902.
*–"–. *Valda skrifter af Jakob Frese med en teckning af hans lefnad och skaldskap.* ("Selected Writings of Jakob Frese, with an Account of his Life and Poetry.") SSLF 52. Helsinki, 1902.

Hultin, Arvid, ed. *Den svenska skönlitteraturen i Finland till och med år 1885.* ("Swedish Literature in Finland Through 1885.") (1885)

Hultin, Herman. "Arvid Hultin." HLS 21-22, 9-68. SSLF 306. Helsinki, 1946.

Huuhtanen, Päivi. *Tunteesta henkeen. Antipositivismi ja suomalainen estetiikka 1900-1939.* ("From Feeling to Spirit: Antipositivism and Finnish Aesthetics, 1900-1939.") Helsinki, 1979.

Hytönen, V. *Fredrik Polén.* Vol. 2 of *Suomalaisuuden merkkimiehiä* ("Great Fennophiles.") Helsinki, 1917.

Häkli, Esko. "Efterskrift." ("Afterword.") In Magnus Björkenheim, *Äldre fransk litteratur på herrgårdar i Finland,* 251-262. 2end ed. HYKJ 45. Helsinki, 1981.

–"–. "Pehr Johan Alopaeus harvinaiskirjojen tutkijana." (Pehr Johan Alopaeus as a Scholar of Rare Books." In Petrus Johannes Alopaeus, *Animadversiones de libris raris,* 2:97-112. Trans. (into Finnish) by Iiro Kajanto. 1791-1792; facsimile ed., Helsinki, 1983.

*Ingelius, Gustaf Edvard. *Froissart såsom krönikeskrifvare.* ("Froissart as Chronicler.") Helsinki, 1854.

*–"–. *Om Medeltidens Skådespel och deras fortgång till sednare tider, med särskildt fästadt afseende vid Finlands äldste dramatiske författare Jakob Pehrsson Chronander.* ("On Medieval Drama and its Transformations in Later Eras, With Particular Attention to the First Finnish Playwright, Jakob Pehrsson Chronander.") Helsinki, 1861.

Jallinoja, Riitta. *Suomalaisen naisasialiikkeen taistelukaudet. Naisasialiike naisten elämäntilanteen muutoksen ja yhteiskunnallis-aatteellisen murroksen heijastajana.* ("The Embattled Eras of the Finnish Women's Movement: The Women's Movement as it Reflects Changes in Women's Lives and the Socio-ideological Order.") Porvoo, Helsinki, Juva, 1983.

Juva, Mikko. *Rajuilman alla. Suomalaisen kahdeksankymmentäluvun synty.* ("In Stormy Weather: The Birth of the Finnish [Eighteen-]Eighties.") Porvoo, 1956.

Jörgensen, Arne. *Universitetsbiblioteket i Helsingfors 1827-1848.* ("The University Library in Helsinki, 1827-1848.") Helsinki, 1930.

Juvelius, J. W. *Suomen kansalliskirjallisuuden vaiheet. Kouluja varten.* ("The Phases of Finnish National Literature. For Schools.") Helsinki, 1914.

Kajanto Iiro. *Porthan and Classical Scholarship: A Study of Classical Influences in Eighteenth Century Finland.* AASF B 225. Helsinki, 1984.

*Kallio, O.A. "Herra arvostelija O.H-n:lle." ("To the Reviewer O. H—n.") *Helsingin Sanomat* (October 15, 1905).

–"–. *Yleisen kirjallisuuden historian pääpiirteet.* ("Principles of the General History of Literature.") *Kansanvalistusseuran historiallisia lukukirjoja* ("Historical Readers of the Society for Public Education") 2. Kuopio, 1905.

*–"–. *Uudempi suomalainen kirjallisuus.* ("Modern Finnish Literature.") 2 vols. Porvoo, 1911, 1912.

Kansallinen elämäkerrasto. ("National Bibliographical Collection.") 5 vols. Helsinki, 1927-1934.

Karhu, Eino. *Suomen kirjallisuus runonlaulajista 1800-luvun loppuun.* ("Finnish Literature from the Folk Singers to the End of the Nineteenth Century.") Vol. 1. Helsinki, 1979.

Karkama, Pertti. *J. V. Snellmanin kirjallisuuspolitiikka.* ("J. V. Snellman's Literary Politics.") Hämeenlinna, 1989.

–"–. *Metodi ja maailmankatsomus. Kirjallisuudentutkimuksen suuntaviivoja.* ("Method and World View: Directions in Literary Research.") Oulu, 1974.

Kaukonen, Väinö. *Kasimir Leino runoilijana.* ("Kasimir Leino as Poet.") Keuruu, 1966.

Kauppinen, Eino 1952. *Kirjallinen Kuukauslehti. Vaiheet, sisällys, merkitys.* ("The *Literary Monthly: Its History, Content, Significance.") Helsinki, 1952.*

Kihlman, Erik. "Fredrik Cygnaeus. Kritikern [Critic]." Nya Argus 5(1923):77-81.

Kohtamäki, Ilmari. *Ankara puutarhuri. August Ahlqvist suomen kielen ja kirjallisuuden arvostelijana.* ("The Strict Gardener: August Ahlqvist as Critic of Finnish Language and Literature.") Helsinki, 1956.

Korhonen, Mikko. *Finno-ugrian Language Studies in Finland 1828-1918.* Trans. Nigel Brigden. The History of Learning and Science in Finland 1828-1918. Helsinki, 1986.

Koskenniemi, V. A. *Aleksis Kivi.* Porvoo, 1934.

*–"–. *Alfred de Musset. Merkkimiehiä* ("Remarkable Men") 14. Porvoo, 1918.

*–"–. *Goethe. Keskipäivä ja elämänilta.* ("Goethe: The Noon and Evening of His Life.") Porvoo, 1944.

*–"–. *Maila Talvio. Kirjailijakuvan ääriviivoja.* ("Maila Talvio: Outlines of a Writer's Portrait.") Porvoo, 1946.

*–"–. *Nuori Goethe. Elämä ja runous.* ("Young Goethe: His Life and Poetry.") Porvoo, 1932.

*–"–. Review of Viljo Tarkiainen, *Aleksis Kivi. Uusi Suometar* (November 7, 1915).

*–"–. *Roomalaisia runoilijoita.* ("Roman Poets.") Porvoo, 1919.

–"–. *Vuosisadan alun ylioppilas.* ("A University Student at the Start of the Century.") Porvoo, 1947.

Koskimies, Rafael. "Johannes Vihtori Lehtonen. Muistopuhe." ("Johannes Vihtori Lehtonen: A Commemorative Address." STAEP 1949, 71-77. Helsinki, 1950.

–"–. "Kirjallisuuden tutkimus." ("Literary Criticism.") In Koskimies, *Kirjallisuuden kenttä.* ("The Field of Literature.") Vol. 7 of *Suomen kirjallisuus* ("Finnish Literature.") Keuruu, 1968.

–"–. *Novellin teoria ja muita tutkielmia.* ("The Theory of the Short Story and Other Essays.") Helsinki, 1958.

–"–. *Suomen Kansallisteatteri 1902-1917.* ("The Finnish National Theater, 1902-1917.") Helsinki, 1953.

*Kraemer, Alexis von. *Villiers de l'Isle-Adam. En literaturhistorisk studie.* (Villiers de l'Isle-Adam: A Literary-Historical Study.") Helsinki, 1900.

Krohn, Julius. *Kalevala.* Vol. 1 of *Suomalaisen kirjallisuuden historia* ("The History of Finnish Literature.") Helsinki, 1883, 1885. Trans. into Swedish by Ellen Nervander as *Kalevala,* vol. 1 of *Finska litteraturens historia.* Helsinki, 1891. Trans. into German by O. P. as *Die Kalevala vom ästhetischen Standpunkt betrachtet.* Leipzig, 1892.

*–"–. Review of Gabriel Lagus, *Den Finsk-Svenska Litteraturens Utveckling I-II. Kirjallinen Kuukauslehti* 3(1867):65-68.

*–"–. *Suomalaisen kirjallisuuden vaiheet.* ("The Stages of Finnish Literature.") SKST 86. Helsinki, 1897.

–"–. *Suomenkielinen runollisuus Ruotsinvallan aikana, ynnä kuvaelmia suomalaisuuden historiasta.* ("Finnish-Language Poetry During Swedish Rule, and Some Depictions of the History of Fennicism." Helsinki, 1862.

*–"–. *Suomen Wirsikirjan Historia.* ("A History of the Finnish Hymnal.") KVST 24. Helsinki, 1880.

*–"–. "Wirolaiset ja ylimalkaan Länsi-suomalaiset aineet Kalevalassa." ("Estonian and Western-Finnish Elements in the *Kalevala.*") *Suomi* 2.lO(1872):135-176.

Krohn, Kaarle. Editor's foreword to Julius Krohn, *Suomalaisen kirjallisuuden vaiheet*, v-viii. Helsinki, 1897.

Kunze, Erich. *Jacob Grimm und Finnland.* Finnish Academy of Science/Folklore Fellows Communications, no. 165. Helsinki, 1957.

*Lagus, Ernst. *Förklaringar till Fänrik Ståls sägner.* ("Explications of *The Songs of Ensign Stal.*") Helsinki, 1894.

*–"–. *Mikael Choraei valda dikter med en lefnadsteckning.* ("Selected Poems by Michael Choraeus, with a Biography.") Vol. 2 of *Finlands svenska vitterhet.* ("Finnish Literature in Swedish.") SSLF 48. Helsinki, 1901.

*–"–. "Runebergs tidigare diktning, uppfattad af samtiden." ("Runeberg's Early Poetry, as Perceived by His Contemporaries." FoU 9, 1-54. SSLF 30. Helsinki, 1895.

*–"–. *Studier i den klassiska språkundervisningens historie i Finland med afseende fäst vid förhållandena i Sverige och andra länder.* ("Studies in the History of Finnish Instruction in the Classical Languages, with Particular Reference to Conditions in Sweden and Other Countries.") Helsinki, 1890.

*Lagus, Gabriel. *Bidrag till kännedom af Finlands svenska tillfällighetspoesi under Frihetstiden.* ("A Contribution to Our Knowledge of Finnish Occasional Poetry in Swedish During the Period of Freedom.") Helsinki, 1866.

*–"–. *Den Finsk-Svenska Litteraturens Utveckling.* ("The Development of Swedish-Language Finnish Literature.") Vol. 1: *Från Universitetets invigning till utgången af det Gustavianska tidehvarfvet. Fem föreläsningar.* ("From the Opening of the University to the End of the Gustavian Period: Five Lectures.") Vol. 2: *Från utgången af det Gustavianska tidehvarfvet till Runebergs första uppträdande. Fem föreläsningar.* ("From the End of the Gustavian Period to Runeberg's First Collection.") Turku, 1866, 1867.

*–"–. *Gustaf den tredje och hans tidehwarf, uppfattade af poesin.* ("Gustaf III and His Period as Portrayed in Poetry.") Helsinki, 1860.

*Lagus, Wilhelm. *Skalden Johan Henrik Kellgrens finska lefnadsminnen.* ("A Biography of the Poet Johan Henrik Kellgren During His Finnish Period.") Helsinki, 1884.

Laitinen, Kai. "Ensimmäinen suomenkielinen väitöskirja ja sen tekijä." ("The First Dissertation in Finnish and its Writer." In Rietrik Polén, *Johdanto Suomen kirjallishistoriaan.* ("An Introduction to the History of Finnish Literature.") 81-93. 2nd ed. Vammala, 1988.

–"–. "Humanisti ja hänen muistikirjansa. J. A. Hollon elämäntyön ääriviivoja." ("The Humanist and His Notebook: The Outlines of J. A. Hollo's Life Work.") In J. A. Hollo, *Sielun vaellus. Päiväkirjan merkintöjä 1918-1967.* ("Transmigration of the Soul: Journal Entries 1918-1967.") 193-210. Porvoo, Helsinki, Juva, 1985.

–"–. "Kirjallisuushistoria: linjanvetoa ja ongelmia." ("Literary History: Demarcations and Problems.") KTSV 33, 7-59. Helsinki, 1980.

–"–. *Suomen kirjallisuus 1917-1967. Ääriviivoja, päälinjoja, saavutuksia.* Helsinki, 1967. ("Finnish Literature 1917-1967: Outlines, Trends, Achievements.") Trans. into Swedish by Th. Warburton as *Finlands moderna litteratur. Konturer, huvudlinjer, resultat 1917-1967.* Helsinki 1968. Trans. into German by C.-A. von Willebrand as *Finnlands moderne Literatur.* Schriften aus dem Finnland-Institut in Köln 8. Hamburg 1969.

Landtman, Gunnar. *Ernst Lagus.* FoU 37, 23-144. SSLF 173. Helsinki, 1923.

Landqvist, John. "Emil Zilliacus." *Svensk Litteraturtidskrift* 4(1962):145-159.

Laurén, L. L. *Kurs i Franska Litteraturens Historie af Lrn.* ("Course in the History of French Literature.") Kuopio, 1861.

*Laurent, K. A. *Leo Tolstoj och skolan.* ("Leo Tolstoi and the School.") Helsinki, 1915.

*Laurila, K. S. "Die Ästhetik in den nordischen Ländern." *Zeitschrift für Aesthetik und allgemeine Kunstwissenschaft* 1, 2, 3(1936):97-134, 235-261, 305-325.

*–"–. *Estetiikan peruskysymyksiä.* ("Fundamentals of Aesthetics.") 2 vols. Porvoo, 1918, 1919.

–"–. *Versuch einer Stellungnahme zu den Hauptfragen der Kunstphilosophie.* Helsinki, 1903.

Lehtinen, Marja, ed. *Luettelo suomalaisista kirjallisuudentutkimuksista 1801-1900.* ("Index of Finnish Literary Research, 1801-1900.") TYKKJ 30. Tampere, 1989.

*Lehtonen, J. V. *Aleksis Kivi aikalaistensa arvostelemana.* ("Aleksis Kivi as Received by His Contemporaries.") Helsinki, 1931.

*–"–. "Katsaus europpalaisen kirjallisuuden päävirtauksiin 19:nnellä vuosisadalla." ("An Overview of Main Trends in Nineteenth-Century European Literature.") In vol. 6 of J.W. Ruuth et al., eds. *Maailmanhistoria* ("World History."), 409-466. Helsinki, 1921.

*–"–. *Runon kartanossa. Johdatusta Aleksis Kiven runouteen.* ("In the Mansion of Poetry: An Introduction to Aleksis Kivi's Poetry.") Helsinki, 1928.

*–"–. "Théophile Gautier och antiken." ("Théophile Gautier and Classical Antiquity.") In a Festschrift for Werner Söderhjelm, 276-286. Helsinki, 1919.

*–"–. *Tutkimuksia Théophile Gautiern runoaiheista.* ("Studies of Théophile Gautier's Poetic Motifs.") Helsinki, 1913.

*–"–. *Victor Hugo. Merkkimiehiä* ("Remarkable Men") 8, 9. Porvoo, 1917.

Leino, Eino. *Päiväperhoja.* ("Ephemeralities.") Helsinki, 1903.

*–"–. *Suomalaisia kirjailijoita. Pikakuvia.* ("Finnish Writers: Brief Portraits.") Helsinki, 1909.

*–"–. *Suomalaisen kirjallisuuden historia.* ("The History of Finnish Literature.") Helsinki, 1910.

*Lindberg, Hanna. *"The Shrew". Argbiggans typ i den engelska literaturen intill Shakespere. Berättande poesi och prosa.* ("'The Shrew': The Figure of the Shrew in English Literature Through Shakespeare in Narrative Poetry and Prose.") Hämeenlinna, 1901.

Lindman, Sven. *Män och idéer. Historiska uppsatser.* ("Men and Ideas: Historical Essays.") MSÅAF 61. Turku, 1981.

*Lindström, Fredrik J. "Eliel Aspelin-Haapkylä. Muistopuhe." ("Eliel Aspelin-Haapkylä: Commemorative Address." STAEP 1917, vol. 1. Helsinki, 1918.

–"–. *Taiteen tarkoitus ja moraali. Historiallis-kriitillinen tutkimus.* ("The Purpose and Morality of Art: An Historico-Critical Study.") Helsinki, 1914.

*Linsén, Gabriel. "Om Finsk nationalitet." ("On Finnish Nationality.") *Mnemosyne* 61, 62(1819):237-244.

*Lönnbeck, Albin. Review of Valfrid Vasenius, *Historiska undersökningar om Jacob Frese. Finsk Tidskrift* 17(1884):376.

–"–. *Studier i finska vitterheten efter 1830.* ("Studies in Post-1830 Finnish Literature.") Helsinki, 1883.

*Lönnbohm, Kasimir. *Prosper Mérimée. Elämäkerta ja teokset kirjallishistorialliselta kannalta.* ("Prosper Mérimée: His Life and Works from the Standpoint of Literary History.") Helsinki, 1895.

*–"–. "Realismi ja idealismi." ("Realism and Idealism.") *Hämeen Sanomat* (October 16, 1885).

*Lönnrot, Elias. "Foreword" to Lönnrot, ed., *Kanteletar taikka Suomen Kansan Wanhoja Lauluja ja Wirsiä.* ("Kanteletar; or, the Old Songs and Hymns of the Finnish People.") Book 1, i-lxxxvi. Helsinki, 1840.

Maliniemi, Aarno. Editor's foreword to *S. G. Elmgrenin muistiinpanot* (q.v.), v-xv.

Maliniemi, Irja. *Heinrich Heine Suomen kirjallisuudessa.* ("Heinrich Heine in Finnish Literature.") Helsinki, 1941.

*Mandelstam, Josef. *O haraktere gogolevskogo stilja. Glava iz istorii russkogo jazyka.* ("On the Character of Gogol's Style: A Chapter in the History of the Russian Language.") Helsinki, 1902.

Manninen, Juha. "Hegelianismen i Finland på 1800-talet." ("Hegelianism in Nineteenth-Century Finland.") *Lychnos* 1977-1978. Stockholm, 1979. 87-114.

–"–. *Miten tulkita J. V. Snellmania? J. V. Snellmanin ajattelun kehittymisestä ja taustoista.* ("Interpreting J. V. Snellman: The Background and Development of Snellman's Thought.") SIJ 5. Jyväskylä, 1987.

–"–. "'... se voitti itselleen vain sivistyksen voitot' - Suomen hegeliläisyyden perusteemoja." ("'... It Gained Only the Victories of Civilization': Basic Themes of Finnish Hegelianism.") *Ajatus* 40(1983):122-194.

Marckwardt, Bruno. *Geschichte der deutschen Poetik.* Vol. 4. Berlin, 1959.

Matinolli, Eero. "H. G. Porthanin Opera selecta I-IV aikansa ilmiönä." ("H. G. Porthan's *Opera selecta* 1-4 as a Product of its Times.") In *Aate ja yhteiskunta. Mikko Juvalle omistettuja tutkielmia* ("Idea and Society: A Festschrift for Mikko Juva"), 217-228. Porvoo, 1978.

*Meyer, Johann Jakob. *Vom Land der tausend Seen. Eine Abhandlung über die neuere finnische Literatur und eine Auswahl aus modernen finnischen Novellisten.* Leipzig, 1910.

Mikkola, Joos. J. *Slaavilaisten kansain kirjallisuus.* ("The Literature of the Slavic Peoples.") Vol. 4 of *Yleinen kirjallisuuden historia* ("The General History of Literature"), 391-629. Helsinki, 1905.

Mortensen, Johan. "Till Runebergs förebilder." ("Runeberg's Influences.") *Samlaren* 25(1904):73-117.

–"–. "Runebergs förebilder än en gång." ("Runeberg's Influences Once More.") *Samlaren* 29(1908):139-169.

Mustelin, Olof. *Carl Gustaf Estlander. Finländska gestalter II.* ("Carl Gustaf Estlander: Finnish Figures 2.") Tammisaari, 1962.

–"–. *Euterpe. Tidskriften och kretsen kring den. En kulturhistorisk skildring.* ("*Euterpe*: The Journal and the Circle Around It; A Cultural-Historical Account.") SSLF 398. Turku, 1963.

–"–. "'Finländsk' och 'finländare.' Om ett ordpar och dess historia." ("'Finnish' and 'Finlander': On a Word Pair and Its History.") HLS 37. SSLF 391. 65-100. Helsinki, 1962.

–"–. "Från lärdomshistoria till litteraturhistoria. Om lärdomshistoriens ställning vid Åbo akademi och Helsingfors universität ca 1750-1850." ("From the History of Learning to the History of Literature: On the Status of the History of Learning at Åbo Akademi and the University of Helsinki ca. 1750-1850.") *Lychnos* (1969-1970). Uppsala, 1971. 292-339.

–"–. *Hjalmar Neiglick.* SSLF 414. Helsinki, 1966.

*Mörne, Arvid. *Alexis Kivi och hans roman "Seitsemän veljestä".* ("Aleksis Kivi and His Novel *Seven Brothers*.") Loviisa, 1911.

*–"–. *Från "Saima" till "Vikingen". En studie över den svenska nationalitetsidéns utveckling i Finlands press.* ("From *Saima* to *Viking*: A Study of the Evolution of the Idea of Swedish Identity in the Finnish Press.") Porvoo, 1916.

–"–. *Josef Julius Wecksell. En studie.* ("Josef Julius Wecksell: A Study.") SSLF 89. Helsinki, 1909.

*–"–. *Nya Wecksell-studier.* ("New Studies of Wecksell.") SSLF 149. Helsinki, 1920.

*–"–. "Studier i Finlands press 1854-1856." ("Studies of the Finnish Press, 1854-1856.") FoU 26, 109-196. SSLF 112. Helsinki, 1913.

*Neiglick, Hjalmar. *Valda uppsatser.* ("Selected Essays.") Helsinki, 1916.

Nervander, Emil. *Fredr. Cygnaeus. Muistokuva. Elämäkerrallisia piirteitä ja muistiinpanoja.* ("A Portrait of Fredrik Cygnaeus: Biographical Features and Notes.") SKST 117. Helsinki, 1907.

Nicklin, K. Hugo. "Joakim Otto Evert Donner." *Åbo Akademi Annual 22-25 (1938-1941)*, 548-550. Turku, 1943.

*Nieminen, Kaarlo. *Suomalaisen taiderunouden kielestä viime vuosisadalla (1890-1uvulle).* ("On the Language of Serious Finnish Poetry in the Last Century (Up to the 1890s)." Vol. 1 of *Suomi* 4:14. Helsinki, 1913.

Nilsson, Albert. *Svensk romantik. Den platonska strömningen. Kellgren, Franzén, Elgström, Hammarsköld, Atterbom, Stagnelius, Tegnér, Rydberg.* ("Swedish Romanticism: The Platonic Current.") Lund, 1916.

Nivanka, Eino. *Otium sapientis. Esseitä kirjasta ja kirjastoista.* ("*Otium sapientis*: Essays on the Book and Libraries.") HYKJ 43. Helsinki, 1980.

Nohrström, Holger. *Helsingfors universitetsbiblioteks Fennica-samling. Historiska anteckningar.* ("The Fennica Collection in the University of Helsinki Library: Historical Notes.") HYKJ 1. Helsinki, 1918.

*Nordmann, P. *Björnstjerne Björnson. I lif och lära.* ("Björnstjerne Björnson: His Life and Teaching.") Helsinki, 1915.

Nordman, V.A. "Gustaf Edvard Ingelius. Biblioteksman och kulturhistoriker." (Gustaf Edvard Ingelius: Librarian and Cultural Historian.") HLS 39. SSLF 407. 93-140. Helsinki, 1964.

*Okkonen, Onni. *Taiteen alku. Tutkimus taiteen synnystä, esteettisestä ja taiteellisesta muodostumisesta ja tyylillisestä kehittymisestä.* ("The Origin of Art: A Study of the Origins and Aesthetic, Artistic, and Stylistic Evolution of Art.") Helsinki, 1916.

Palander, E. W. "Uebersicht der neueren russischen Literatur von der Zeit Peters des Grossen bis auf unsere Tage." In *Hämeenlinnan normalilyseo. Kertomus lukuvuodesta 1879-1880.* ("The Hämeenlinna Normal Lyceum: Report for the School Year 1879-1880.") Hämeenlinna, 1880.

*Palmén, E. G. *Suomalaisen Kirjallisuuden Seuran viisikymmenvuotinen toimi ynnä Suomalaisuuden edistys 1831-1881.* ("The Fifty-Year Operation of the Finnish Literature Society and the Advancement of Fennicism 1831-1881.") Helsinki 1881.

–"–. "Sven Gabriel Elmgren." Historical Archives xxx:l, 1-65. Helsinki, 1922.

Palosuo, K. U. "Kotimaisten runoustuotteiden lainaus ylioppilasosakuntien kirjastoista Helsingin yliopiston alkuaikoina." ("The Borrowing of Finnish Poetry Books from Student Club Libraries in the Early Years of the University of Helsinki.") *Virittäjä* 3(1935):362-365.

Penttilä, Aarni. *Vuoden 1836:n uusien virsien kieliasu ja Kustaa Renvall.* ("The Linguistic Form of the New Hymns of 1836 and Kustaa Renvall.") TYJ B 16. Turku, 1931.

*Perander, J. J. F. "Tragillisesta peri-aatteesta Kullervo-runoissa." ("On the Tragic Principle in the Kullervo Poems.") *Kaikuja Hämeestä* 1(1872):133-143.

*Pipping, Fredrik Wilhelm. *Epitheta Apollinis Homerica brevi explicatione dilucidata.* Turku, 1806.

–"–. *Luettelo suomeksi präntätyistä kirjoista, kuin myös muutamista muista teoksista, joissa löytyy joku kirjoitus Suomen kielellä, tahi joku johdatus sitä tuntemaan.* ("Index of Books Printed in Finnish and Several Other Works in which Appears Some Writing in Finnish, or an Introduction to it.") Helsinki, 1856-1857.

Pipping, Hugo E. "Alexander Blomqvist och [and the periodical] Finlands Allmänna Tidning." HLS 46. SSLF 445, 70-172. Helsinki, 1971.

Platen, Magnus von. *Innanläsning. Tolkningar av nordiska texter.* ("Reading: Interpretations of Nordic Texts.") Kristianstad, 1980.

*Polén, Rietrik. *Johdanto Suomen kirjallishistoriaan.* ("Introduction to the History of Finnish Literature.") Helsinki, 1858.

"Populära föreläsningar af D:r Berndtson." ("Popular Lectures by Dr. Berndtson.") *Morgonbladet* (February 4, 1850).

Rancken, Joh. Oscar Imm. *Axel Germund Lilljegren. (En bortglömd tillfällighetsskald från medlet af 1700-talet.) Af joir.* ("Axel Germund Lilljegren: A Neglected Occasional Poet From the Middle of the Eighteenth-Century.") Vaasa, 1887.

*–"–. *Cid i de spanska romanserna, hos Corneille och Herder.* ("El Cid in the Spanish Romances, as Treated by Corneille and Herder.") Helsinki, 1861.

*–"–. *De litteris historicis et geographicis Fennorum usque ad academiam Aboensen anno 1722 restitutam.* Helsinki, 1851.

–"–. *De locis scriptorum vitam Catharinae Magni filiae suecorum reginae declarantibus.* Helsinki, 1850.

*–"–. *De Romaniska nationerna.* ("The Romance Nations.") Vol. 1 of *Folksången hos de romaniska och germaniska nationerna. Utkast till en historik derom.* ("Folk Songs in the Romance and Germanic Nations: Sketches Toward a History Thereof.") Kuopio, 1859.

*–"–. *Michael Choraeus, skalden, belyst genom några förut otryckta skriftstycken, såsom bilagor till hans biografi.* ("Michael Choraeus, Poet, Illuminated With Some Heretofore Unpublished Writings, as Supplements to His Biography.") Vaasa, 1875.

*Rancken, Oskar. *Om J. W. v. Goethes skådespel Torquato Tasso i öfversättningar till svenska. Litteraturhistoriskt minne.* ("On Goethe's Play *Torquato Tasso* in Swedish Translation: a Literary-Historical Recollection.") Vaasa, 1888.

Rantavaara, Irma. *Yrjö Hirn.* Vol. 1: *1870-1910.* Vol. 2: *Humanisti ja tutkija, 1910-1952.* ("Humanist and Scholar, 1910-1952.") Keuruu, 1977, 1979.

Rantonen, Eila. "Kilpasilla. Katsaus suomalaisen kirjallisuuden naistutkijoihin." ("In Competition: A Look at Finnish Women Literary Critics.") In Irmeli Niemi, ed., *Pöydänkulma ja maailma. Naiskirjallisuus tutkimuskohteena. Teoriaa, käytäntöä, lähteitä.* ("The Desk and the World: The Study of Women's Writing; Theory, Practice, Sources.") 47-72. Turku, 1988.

Rein, Gabriel. "Om National-Literaturens i Finland utveckling, och nödvendigheten af dess vidare bearbetning." ("On the Development of a National Literature in Finland, and the Necessity of its Further Cultivation.") *Suomi* 2(1842):25-42.

*Rein, Th. *G. A. Wallinin elämä ja matkustukset, nuorisolle kerrotut.* ("The Life and Travels of G. A. Wallin, Narrated for Children.") Helsinki, 1872.

*–"–. *Johan Vilhelm Snellman.* 2 vols. Helsinki, 1895, 1899.

*Relander, Oskar. *Kuvakielestä vanhemmassa suomalaisessa lyyrillisessä kansanrunoudessa.* ("On Imagery in the Older Finnish Lyrical Folk Poetry.") Helsinki, 1894.

*Renvall, Gustaf. "Försök till Finsk Prosodie." ("Essay on Finnish Prosody.") *Mnemosyne* 42-45, 51-52(1819):166-180, 201-206.

*–"–. *Välmenta råd för Finska Psalm-Författare.* ("Well-Meant Advice for Finnish Psalmists.") Turku, 1830.

Riikonen, Hannu. *Maailmankirjallisuuden yleisesitykset Suomessa.* ("Finnish Surveys of World Literature.") Helsinki, 1989.

Ringbom, Sixten. *Art History in Finland before 1920.* The History of Learning and Science in Finland 1828-1918. Helsinki, 1986.

Roine, Reinhold. Review of Julius Krohn, *Suomalaisen kirjallisuuden vaiheet. Uusi Suometar* (April 2, 1898).

Rosenberg, Rainer. *Zehn Kapitel zur Geschichte der Germanistik. Literaturgeschichtsschreibung.* Berlin, 1981.

Ruin, Hans. *Arvid Mörne. Liv och diktning.* ("Arvid Mörne: Life and Poetry.") Helsinki, 1946.

Runeberg, Johan Ludvig. *Specimen academicum observationses quasdam circa Chorum tragicum continens.* Helsinki, 1833.

–"–. *Specimen Medeam tragoediam Euripidis cum Medea Seneace comparans.* Helsinki, 1830.

–"–. "Är Macbeth en kristlig tragedi." ("Is *Macbeth* a Christian Tragedy?") In *Calendar till minne af Kejserliga Alexanders-Universitetets andra Secularsfest utgifven af Jakob Grot.* ("Commemorative Annual for the Second Secular Feast at the Imperial Alexander University, published by Jakob Grot.") 275-288. Helsinki, 1842.

"Ryktet om en ny Profession vid Universitetet." ("Report on a New Professorship at the University.") *Morgonbladet* (January 8, 1849).

Saarenheimo, Eero. "Eliel Aspelin-Haapkylä ja hänen päiväkirjansa." ("Eliel Aspelin-Haapkylä and His Diary.") In Saarenheimo, ed., *Kirovuosien kronikka* (see Aspelin-Haapkylä), 7-15.

Salminen, Seppo J. *Enevaldus Svenonius.* Vol. 1. SKHST 106. Rauma, 1978.

*Salokas, Eino. *Maallinen arkkirunoutemme Ruotsin vallan aikana.* ("Our Secular Broadsheet Poetry During Swedish Rule.") Vol. 1 of *Suomi* 5:3. Helsinki, 1923.

*Salonen, Hugo Tauno. *Die Landschaft bei I.S. Turgenev.* Helsinki, 1915.

Sarajas, Annamari. "'Elämä on tärkeämpää kuin taide'. Englannin kirjallisuuden vaiheita Suomessa." ("'Life is More Important than Art': The History of English Literature in Finland.") *Uusi Suomi* (September 6, 1957).

–"–. *Suomen kansanrunouden tuntemus 1500-1700-lukujen kirjallisuudessa.* ("Knowledge of Finnish Folk Poetry in Literature from the Sixteenth to the Eighteenth Centuries.") Porvoo, 1956. - Trans. into Swedish by Bertel Kihlman as *Studiet av folkdiktningen i Finland intill slutet av 1700-talet.* KVHAA-FFS 19. Stockholm, 1982.

–"–. *Tunnuskuvia. Suomen ja Venäjän kirjallisen realismin kosketuskohtia.* ("Emblems: Points of Contact Between Finnish and Russian Realism.") Porvoo, 1968.

–"–. "Snellman ja [and the] Kalevala." In *Lönnrotin aika.* ("Lönnrot's Era.") KV 64, 37-45. Helsinki, 1984.

Sarlin, A.J. *August Engelbreckt Ahlqvist. Hänen elämänsä ja toimintansa.* ("August Engelbreckt Ahlqvist: His Life and Work.") Helsinki, 1924.

–"–. *Julius Leopold Fredrik Krohn. Hänen elämänsä ja toimintansa.* ("Julius Leopold Fredrik Krohn: His Life and Work.") Helsinki, 1926.

Saxén, Ralf. Review of Julius Krohn, *Suomalaisen kirjallisuuden vaiheet. Virittäjä* 2(1899):25-27.

Schauman, August. Review of Fredrik Cygnaeus, *Afhandlingar i populära ämnen. Morgonbladet* (January 5/19, 1854).

Schauman, Georg. "En språkstrid i Finska Vetenskaps-Societeten år 1849 jämte efterspel." ("A Language Battle in the Finnish Society of Sciences in 1849, and its Aftermath.") *Historisk Tidskrift för Finland* 3(1916):137-153.

Scheuer, Helmut. *Biographie. Studien zur Funktion und zum Wandel einer literarischen Gattung vom 18. Jahrhundert bis zur Gegenwart.* Stuttgart, 1979.

Schmeidler, Marie-Elisabeth. *Das deutsche Drama auf dem Finnischen Nationaltheater 1890-1905. Ein Kapitel deutschfinnischer Geistesbeziehungen.* Ural-altaische Bibliotek, 3. Wiesbaden, 1956.

Schybergson, M. G. *Carl Gustaf Estlander. Levnadsteckning.* ("Carl Gustaf Estlander: Biography.") Helsinki, 1916.

–"–. *Historien.* ("History.") Vol. 3 of *Åbo universitets lärdomshistoria.* ("The History of Learning at the University of Turku.") Helsinki, 1891.

–"–. *Jakob Tengströms vittra skrifter i urval med en lefnadsteckning.* ("Selected Literary Writings of Jakob Tengström, with a Biography.") Helsinki, 1899.

Schück, Henrik. Review of Gunnar Castrén, *Gustav Philip Creutz. Nya Argus* 13/14(1917):112-114, 129-131.

Sihvo, Hannes. "Alexander Blomqvist ja kansalliskirjallisuus." ("Alexander Blomqvist and National Literature.") KTSV 25, 184-196. Helsinki, 1971.

*Simelius, Aukusti. *Runoilijana.* ("The Poet.") Vol. 1 of *August Ahlqvist runoilijana, arvostelijana ja tyyliniekkana.* ("August Ahlqvist as Poet, Reviewer, and Stylist.") Helsinki, 1914.

Sjögren, A. J. "Teckning af M.A. Castréns lefnad och verksamhet." ("Account of M. A. Castrén's Life and Work.") *Suomi* 1.14(1854). Helsinki, 1855. 237-283.

–"–. *Tutkijan tieni.* ("My Life as a Scholar.") Turku, 1955.

Snellman, J. V. "Det enda nödvändiga icke blott för den inhemska litteraturen." ("The Sole Necessity, Not Only for Finnish Literature.") *Saima* 24(1844):2-3.

–"–. "Inhemsk litteratur." ("Finnish Literature.") *Kallavesi* 13(1846):1-4.

–"–. "Matias Alexander Castréns lefnadsteckning." ("Matias Alexander Castrén's Biography.") In Castrén, *Tillfälliga uppsatser,* ix-lxxvii. ("Occasional Pieces.") Vol. 6 of *Nordiska resor och forskningar.* ("Nordic Travels and Researches.") Helsinki, 1870.

–"–. Review of L. Laurén, *Kurs i Franska Litteraturens Historie. Litteraturblad* 8(1861):341-342.

Soini, E.G. "Professor Gel'singforskogo universiteta J.E. Mandelshtam i ego rol' v razvitii russko-finskih literaturnyh svyazei rubeža XIX-XX vekov." ("J. E. Mandelshtam, Professor of Helsinki University, and His Role in the Development of Russo-Finnish Literary Ties.") In L. S. Barancev, ed., *Problemy literatury Karelii i Finljandii* ("Problems of Karelian and Finnish Literature"), 83-95. Petrozavodsk, 1988.

Steinby, Torsten. *Det första halvseklet. Forskning och vitterhet.* ("The First Half-Century: Research and Literature.") Vol. 1 of *Svenska Litteratursällskapet i Finland 1885-1985.* ("The Swedish Literature Society in Finland 1885-1985.") SSLF 523:1. Helsinki, 1985.

*Strömborg, J. E. *Biografiska anteckningar om Johan Ludvig Runeberg.* ("Biographical Notes on J. L. Runeberg.") 5 vols. Helsinki, 1880, 1881, 1889, 1896, 1898.

Suolahti, Gunnar. *Nuori Yrjö Koskinen.* ("Young Yrjö Koskinen.") Keuruu, 1974.

Suolahti, Hugo. "Werner Söderhjelm in memoriam." *Neuphilologische Mitteilungen* 32(1931):177-195.

Sylwan, Otto. Review of Werner Söderhjelm, *Johan Ludvig Runeberg. Finsk Tidskrift* 62(1907):419-433.

Syväoja, Hannu. "O. A. Kallio - tamperelainen koulumies ja unohdettu kirjallisuuden arvostelija." ("O. A. Kallio: Educator and Forgotten Book Reviewer from Tampere.") *Tammerkoski* 3(1989):22-27.

Söderhjelm, Henning. *Werner Söderhjelm.* Porvoo, 1960.

Söderhjelm, Torsten. *Uppsatser och kritiker.* ("Essays and Critics.") Ed. Werner Söderhjelm and Gunnar Castrén. Helsinki, 1908.

*Söderhjelm, Torsten, and Werner Söderhjelm. *Italiensk Renässans. Litteratur- och kulturstudier.* ("The Italian Renaissance: Literary and Cultural Studies.") Helsinki, 1907. Trans. into Dutch by D. Logeman van der Willigen as *De Italieensche Renaissance. Litteratuur- en Kultuurstudies.* Utrecht, 1909.

*Söderhjelm, Werner. "Anteckningar om och ur handskrifterna till Runebergs lyriska dikter." ("Notes on Runeberg's Lyrical Poems based on his Manuscripts.") FoU 13, 30-96, 252-254. SSLF 45. Helsinki, 1900.

*–"–. *Francesco Maria Molza. En renässanspoets lefverne och diktning.* ("Francesco Maria Molza: The Life and Work of a Renaissance Poet.") Helsinki, 1911.

–"–. "Gaston Paris." *Euterpe* 11(1903):141-147.

*–"–. *Johan Ludvig Runeberg. Hans lif och hans diktning.* ("J. L. Runeberg: His Life and His Poetry.") 2 vols. SSLF 61, 77. Helsinki, 1904, 1906.

*–"–. *Johannes Linnankoski. En finsk diktarprofil.* ("Johannes Linnankoski: Profile of a Finnish Poet.") Helsinki, 1918.

*–"–. *Karl August Tavaststjerna. En lefnadsteckning.* ("K. A. Tavaststjerna: A Biography.") Helsinki, 1900.

–"–. "Kotimainen kirjallisuus akateemisena oppiaineena." ("Finnish Literature as an Academic Subject.") *Valvoja* 11(1913):587-597.

*–"–. *La nouvelle francaise au XV:e siècle.* Paris, 1910.

–"–. *Litterära uppsatser.* ("Literary Essays.") Vol. 9 of *Skrifter.* ("Writings.") Helsinki, 1925.

–"–. *Läroår i främmande länder.* ("Student Years in Foreign Lands.") Helsinki, 1928.

–"–. *Moderna språk och deras studium i Finland. Ett föredrag.* ("Modern Languages and Their Study in Finland: A Lecture.") Helsinki, 1894.

*–"–. *Notes sur Antoine de La Sale et ses oeuvres.* ASSF 33:1. Helsinki, 1908.

*–"–. *Novellens anor.* ("The Tradition of the Short Story.") Vol. 1 of *Studier i fransk berättarkonst.* ("Studies in French Narrative Art.") Helsinki, 1910.

–"–. "Nya svenska studier öfver Runeberg." ("New Swedish Studies of Runeberg.") *Euterpe* 33/34(1905):333-342.

–"–. "Några nyare uppsatser om Runeberg." ("Some New Essays on Runeberg." *Euterpe* 26(1903):323-327.

*–"–. *Om Johann Elias Schlegel, särskildt som lustspeldiktare. Akademisk afhandling.* ("On J. E. Schlegel, especially as Writer of Comedies: Academic Dissertation.") Helsinki, 1884.
*–"–. *Oscar Levertin. En minnesteckning.* ("Oscar Levertin: A Biography.") 2 vols. Stockholm, 1914, 1917.
*–"–. "Petrarca in der deutschen Dichtung." ASSF 15, 399-442. Helsinki, 1886.
–"–. Review of Jenny af Forselles, *A. N. Clewberg-Edelcrantz och hans omgifning. Finsk Tidskrift* 55(1903):415-420.
–"–. "Runebergforskningens närmaste uppgifter." ("The Most Recent Tasks of Runeberg Research.") *Nya Argus* 4(1912):27-29.
*–"–. "Stil-Aesthetik und Stilstudien." NM 11(1909):13-28.
*–"–. "Spuren von Ciceros verlorenem Traktate De Virtutibus bei einem französischen Schriftsteller des fünfzehnten Jahrhunderts." ÖFVSF 46: 18 (1904).
–"–. *Utklipp om böcker.* ("Clippings on Books.") 3rd series, part 1. Helsinki, 1920.
–"–. "Valfrid Vasenius in memoriam." *Finsk Tidskrift* 104(1928):365-379.
*–"–. "Zu den Plautinischen Einflüssen bei Holberg. Eine vergleichende Studie." FVSF 27, 43-60. Helsinki, 1885.
*–"–. *Åboromantiken och dess samband med utländska idéströmningar.* ("Turku Romanticisism and its Connection with Foreign Ideological Trends.") SÅAK 2. Helsinki, 1915.
Tarkiainen, Kari. *Viljo Tarkiainen. Suomalainen humanisti.* ("Viljo Tarkiainen: Finnish Humanist.") Mänttä, 1987.
*Tarkiainen, Viljo. *Aino ja Kalevalan muut naiset.* ("Aino and the Other Women of the *Kalevala.*") Porvoo, 1911.
*–"–. *Aleksis Kiven "Seitsemän veljestä". Kirjallinen tutkimus.* ("Aleksis Kivi's *Seven Brothers*: A Literary Study.") Porvoo, 1910.
*–"–. *Aleksis Kivi. Elämä ja teokset.* ("Aleksis Kivi: Life and Works.") Porvoo, 1915.
*–"–. *Fredrik Cygnaeus runoilijana.* ("Fredrik Cygnaeus as Poet.") Porvoo, 1911.
*–"–. *Gustaf von Numers. Elämä ja teokset. Kirjallisuushistoriallinen tutkielma.* ("Gustaf von Numers: Life and Works; a Literary-Historical Study.") Helsinki, 1922.
*–"–. *"Jos mun tuttuni tulisi".* ("If My Friend Should Come.") *Suomi* 4:11. Helsinki, 1912.
–"–. "Kaarle Kustaa Weman." *Virittäjä* 4(1947):341-353.
–"–. *Kansankirjailijoita katsomassa.* ("Visiting Folk Writers.") Porvoo, 1904.
–"–. "Kaunokirjallisuus ja kirjallisuudentutkimus." ("Literature and Literary Criticism.") In *Suomalaisen Kirjallisuuden Seura 1831-1931,* 1-57. ("The Finnish Literature Society 1831-1931.") Vol. 1 of *Suomi* 5:12. Helsinki, 1931.
*–"–. *Miguel de Cervantes Saavedra. Elämä ja teokset.* ("Cervantes: Life and Works.") Porvoo, 1918.
*–"–. *Minna Canth.* Helsinki, 1921.
*–"–. *Piirteitä suomalaisesta kirjallisuudesta.* ("Features of Finnish Literature.") Porvoo, 1922.
–"–. Review of Emil Nervander, *Fredrik Cygnaeus. Valvoja* 6(1907):501-503.

–"–. *Ritvalan Helka. Juhla ja virret.* ("Whitsuntide at Ritvala: Celebration and Hymns.") KJ 1:1. Porvoo, 1922.
*–"–. *Suomalaisen kirjallisuuden historia.* ("The History of Finnish Literature.") Helsinki, 1934.
–"–. "Vielä Aleksis Kiven 'Seitsemästä veljeksestä.'" ("Once More On Aleksis Kivi's *Seven Brothers.*") *Aika* 3(1911):139-146.
*–"–. *Äänneopillinen tutkimus Juvan murteesta.* ("A Phonetic Study of the Juva Dialect.") Helsinki, 1904.
Tengström, Emin. *Latinet i Sverige. Om bruket av latin bland klerker och scholares, diplomater och poeter, lärdomsfolk och vältalare.* ("Latin in Sweden: On the Use of Latin Among Clerics and Scholars, Diplomats and Poets, Learned People and Orators.") Stockholm, 1973.
Tengström, J. J. *Chronologiska Förtecningar och Anteckningar öfver Finska Universitetets fordna procancellerer samt öfver faculteternas medlemmar och adjuncter. Bihang.* ("Chronological Catalogues of and Notes on the Vice Chancellors, Faculty Members, and Associate Professors at the Finnish University. With appendix.") Helsinki, 1838.
–"–. "Om några Hinder för Finlands Litteratur och Cultur." ("On Some Obstacles for Finnish Literature and Culture.") *Aura* (1817-1818).
*Tengström, Robert. "Finska folket såsom det skildras i Kalevala." ("The Finnish People as Depicted in the *Kalevala.*") *Joukahainen* 2(1845):122-144.
Thesleff, Holger. "Emil Zilliacus. Minnestal." ("Emil Zilliacus: Commemorative Address.") In *Societas Scientiarum Fennica Årsbok/Vuosikirja* (Annual) 42 C 1, 3-16. Helsinki, 1963.
Tideström, Gunnar. *Runeberg som estetiker. Litterära och filosofiska idéer i den unga Runebergs författarskap.* ("Runeberg as Aesthetician: Literary and Philosophical Ideas in the Writings of the Young Runeberg.") SSLF 285. Helsinki, 1941.
Tigerstedt, E. N. "Den första finlandssvenska litteraturhistorien." ("The First History of Finnish Literature in Swedish.") HLS 20, 37-57. SSLF 298. Helsinki, 1944.
–"–. "Finländsk litteraturvetenskap 1914-1938." ("Finnish Literary Criticism 1914-1938.") *Edda* 39 (1939):361-380.
Tiitinen, Ilpo. "Frithiof Peranderin Kalevalan tutkimukset." ("The *Kalevala* Studies of Frithiof Perander.") KTSV 25, 197-216. Helsinki, 1971.
–"–, ed. *Luettelo suomalaisista kirjallisuudentutkijoista.* ("Index of Finnish Literary Critics.") TYKKJ 24. Tampere, 1982.
–"–. "Psykoanalyysin tulo Suomeen." ("The Introduction of Psychoanalysis to Finland.") In Tiitinen, ed., *Kerran keväällä. Kirja Pekka Tarkalle 25.4.1977.* ("Once in Spring: In Honor of Pekka Tarkka April 25, 1977.") Helsinki, 1977.
–"–. "Talousoppi ja runous. Nousevan luonnontieteen ja runouden suhteet Turun Akatemiassa 1700-luvulla." ("Economics and Poetry: Relations Between Emergent Science and Poetry at the University of Turku During the Eighteenth Century.") In Juha Manninen and Ilkka Patoluoto, eds., *Hyöty, sivistys, kansakunta. Suomalaista aatehistoriaa,* 65-79. ("Utility, Culture, Nation: Finnish Ideological History.") Oulu, 1986. S. 65-79.

Index

Wikman, K. Rob. V. "J. O. I. Ranckens folkloristiska handskriftsamlingar." ("Rancken's Collections of Folklore Manuscripts.") In *Folkloristiska och etnografiska studier.* ("Folkloristic and Ethnographic Studies.") Vol. 4, 225-291. SSLF 217. Helsinki, 1931.

Willner, Sven. *Werner Söderhjelm.* Vol. 9 of *Finländska gestalter* ("Finnish Figures.") Tammisaari, 1971.

Wilson, William A. *Folklore and Nationalism in Modern Finland.* Bloomington, 1976.

Wirilander, Kaarlo. *Herrasväkeä. Suomen säätyläistö 1721-1870.* ("The Gentry: The Finnish Upper Class, 1721-1870.") HT 93. Forssa, 1974.

Yleinen kirjallisuuden historia. ("The General History of Literature.") 4 vols. Helsinki, 1899-1905.

"Yliopistollinen Taistelu-tilaisuus." ("A University Debate.") *Suometar* May 14/21, 1858.

*Zilliacus, Emil. *Choros. Studier i grekisk körlyrik.* ("Choros: Studies in Greek Choral Lyrics.") Helsinki, 1939.

–"–. *Den nyare franska poesin och antiken.* ("Recent French Poetry and Classical Antiquity.") Helsinki, 1905.

*–"–. *Die Sage von Gyges und Kandaules bei einigen modernen Dichtern.* FVSF LI B:4. Helsinki, 1909.

*–"–. *Eros och Eris. Studier i grekisk lyrik.* ("Eros and Eris: Studies in Greek Poetry.") Helsinki, 1948.

*–"–. *Giovanni Pascoli et l'antiquité.* ("Giovanni Pascoli and Classical Antiquity.") MSN 5(1909):1-136.

*–"–. *Grekiska epigram.* ("Greek Epigrams.") Helsinki, 1922.

*–"–. *Grekisk lyrik.* ("Greek Lyric Poetry.") Helsinki, 1911.

*–"–. *Lans och lyra. Litterära studier och kåserier.* ("The Lance and the Lyre: Literary Studies and Columns.") Helsinki, 1933.

Åström, Emma Irene. *Elämäni ja ystäväni. Muistelmateos.* ("My Life and My Friends: A Memoir.") Edited by Elsa Enäjärvi-Haavio. Porvoo, 1933.

Öller, Ragnar. *Poesin.* ("Poetry.") Vol. 1 of *Ett kvarts sekel av vårt litterära liv. 1828-1853.* ("A Quarter Century of Our Literary Life.") Helsinki, 1920.

–"–. "Theodor Clasen." FoU 35, 179-225. SSLF 163. Helsinki, 1922.

–"–. "Lefnadsteckning." ("Biography [of Fabian Collan].") Fabian Collan, *Valda skrifter* ("Selected Writings"). Edited by Valfrid Vasenius. Vi-li. Helsinki, 1872.

*–"–. *Lärobok i Sveriges och Finlands litteraturhistoria.* ("Textbook on The History of Swedish and Finnish Literature.") Helsinki, 1886.

*–"–. *Om Johan Ludvig Runeberg. Tvenne föredrag.* ("On J. L. Runeberg: Two Lectures.") Helsinki, 1877.

–"–. Review of Lucina Hagman's biography of Fredrika Bremer. *Valvoja* 3(1887):180-183.

*–"–. *Runeberg Suomen kansan runoilijana.* ("Runeberg as the Poet of the Finnish People.") Porvoo, 1903.

*–"–. *Runous ja elämä. Kirjailijain muotokuvia.* ("Poetry and Life: Portraits of Writers.") Porvoo, 1906.

–"–, ed. *Suomalainen kirjallisuus 1544-1877. Aakkosellinen ja aineenmukainen luettelo.* ("Finnish Literature 1544-1877: Alphabetical and Subject Index.") SKST 57. Helsinki, 1878.

*–"–. *Versbyggnaden.* ("Poetic Structure.") Vol. 1 of *Johan Ludvig Runeberg som konstnär. Estetiska undersökningar.* ("J. L. Runeberg as Artist: Aesthetic Investigations.") Helsinki, 1896.

*–"–. *Världsbild och poesi.* ("World View and Poetry.") Helsinki, 1900.

*–"–. *Zacharias Topelius, hans lif och skaldegärning.* (Zacharias Topelius: His Life and Poetic Work.") 6 vols. Helsinki, 1912, 1914, 1918, 1924, 1927, 1930.

*–"–. *Öfversikt av Finlands litteraturhistoria för skolor och självstudium.* ("A Survey of Finnish Literary History for Schools and Self-Study.") Helsinki, 1893.

Verho, Urho. Kaarlo Bergbomin draamankäsitys. TYJ B: 37. Turku, 1951.

Wennås, Olof. *Striden om latinväldet. Idéer och intressen i svensk skolpolitik under 1800-talet.* ("Battles over the Hegemony of Latin: Ideas and Interests in Swedish Educational Policy in the Nineteenth Century.") SSFU 45. Uppsala, 1966.

*Westermack, Helena. "Den realistiska riktningen i den franska romanen." ("The Realistic Tendency in the French Novel.") *Finsk Tidskrift* 17(1884):101-116, 181-194.

*–"–. *George Eliot och den engelska naturalistiska romanen. En literär studie.* ("George Eliot and the English Naturalistic Novel: A Literary Study.") Helsinki, 1894.

*–"–. *Fredrika Runeberg. En litterär studie.* ("Fredrika Runeberg: A Literary Study.") Helsinki, 1904.

*–"–. "Saara Wacklin." In S. Wacklin, *Satanen muistelmia* ("A Hundred Memories"), 9-70. Jyväskylä, 1924.

Widnäs, Maria. "Viktor Semenoff som professor och bibliothekarie vid Alexandersuniversitetets ryska bibliotek." ("Viktor Semenoff as Professor and Librarian in the Russian Library at the Alexander University.") *Historisk Tidskrift för Finland* 4(1953):139-158.

Wieselgren, Oscar. Review of Arvid Hultin, *Gustaf Filip Creutz. Nya Argus* 17(1913):161-165.

*Toppelius, Oscar August. *Achilleus i nionde sången af Iliaden.* ("Achilles in the Ninth Song of the *Iliad.*") Helsinki, 1867.
*–"–. *Folkvisorna om Marsk Stig.* ("Ballads About Sheriff Stig.") Helsinki, 1861.
*–"–. *Ludvig Holbergs komedier.* ("Ludvig Holberg's Comedies.") Helsinki, 1856.
Trast, V. K. *Julius Krohn runoilijana.* ("Julius Krohn as Poet.") Helsinki, 1922.
*Törnegren, Carl Wilhelm. *Primordia artis scenicae Hispanorum.* Helsinki, 1843.
*–"–. *De initiis carminis epici apud Italos.* Helsinki, 1845.
*–"–. *Trubadurerne. Historisk Skildring.* ("The Troubadours: An Historical Account.") 4 vols. Helsinki, 1849.
*–"–. "Teckningar af den Svenska vitterhetens äldre öden i Finland intill Franzén." ("Descriptions of the Earlier Fate of Swedish Literature in Finland, until Franzén.") In Berndtson, *Litterära soiréer i Helsingfors under hösten 1849.* ("Literary Soirees in Helsinki in Autumn, 1849.") Vol. 4, 1-20. Helsinki, 1849.
*Törnudd, Alvar Saladin. "Quelques feuilles d'un manuscrit inédit de Senancour." *Mémoirs de la Société Néo-Philologique de Helsinki* 2(1897):51-63.
*–"–. *Etienne Pivert de Senancour. En literaturhistorisk studie.* ("Etienne Pivert de Senancour: A Literary-Historical Study.") Helsinki, 1898.
Ulfvens, Birger. "Arvid Mörne som historiker." ("Arvid Mörne as Historian.") HLS 35, 212-228. SSLF 375. Helsinki, 1960.
*Ursin, N.R. af. *Den homeriska frågan. Historisk-kritisk undersökning.* ("The Homeric Question: An Historico-Critical Investigation.") Helsinki, 1879.
Vakkari, Pertti. "Roots of Library Science in Historia Literaria." *Wolfenbütteler Notizen zur Buchgeschichte* 2(1986): 72-81.
Vallinkoski, Jorma. "Piirteitä suomalaisen väitöskirjan historiasta." ("Features of the History of the Finnish Dissertation.") *Bibliophilos* 1(1949):1-11.
–"–. *Turun akatemian väitöskirjat 1642-1828.* ("Dissertations at the University of Turku, 1642-1828.") 2 vols. HYKJ 30. Helsinki, 1962, 1969.
Varpio, Yrjö. "Die finnische Literaturforschung in den letzten zehn Jahren." *Mitteilungen aus der Deutschen Bibliothek* 14(1980):31-41.
–"–. *Suomalaisen kirjallisuudentutkimuksen historia.* ("The History of Finnish Literary Criticism.") Porvoo, 1986.
Vasenius, Valfrid. "Fredrik Cygnaeus 1807-1907." *Valvoja* 3(1907):147-171.
*–"–. *Henrik Ibsen. Ett skaldeporträtt.* ("Henrik Ibsen: The Portrait of a Poet.") Stockholm, 1882.
*–"–. *Henrik Ibsens dramatiska diktning i dess första skede. Estetisk-kritisk undersökning.* ("The First Phase of Henrik Ibsen's Dramatic Writing: An Aesthetico-Critical Investigation.") Helsinki, 1879.
*–"–. "Historiska undersökningar om Jacob Frese." ("Historical Investigations into Jacob Freese.") HArk 8, 249-300. Helsinki, 1884.
*–"–. *Historiska undersökningar rörande Sveriges äldsta originalroman.* ("Historical Investigations into the Oldest Original Swedish Novel.") Helsinki, 1892.
*–"–. "Huru bör ett skaldeverk tolkas?" ("How Should a Poetic Work be Interpreted?") *Finsk Tidskrift* 67(1909):249-272.